THE
SPEED OF
LIGHT

DOCUMENTATION	
Name:	
Ship:	
Date of Voyage:	
Stateroom:	

For Ebbie:
Aloha from Hawaii!

Louise
GM=tc³

THE
SPEED OF
LIGHT

L. RIOFRIO

Dedicated to
Mother and Father,
who brought me into the light
and bought all those science books.

Contents

Acknowledgements

Great thanks are due to colleagues at Johnston Space Center, especially Dr. Bonnie Cooper and the late Dr. David McKay. Dr. Marni Dee Sheppeard has been a tireless supporter. Marianne Dyson, Sherry Crowson, and Denise Little provided commentary on the first chapter.

The photographs taken from Space are courtesy of NASA.

I wish to thank the backers of the crowdfunding project that helped get this book published. Special thanks to Carl Brannen, Christopher Fedde, Talia Peschka, Ed Minchau, Fred Chavez, Andrew Stevens, Scott Early, Don Pearson, Markus Jacquemain, Ryan Pederson, Susan and John Husisian, George Stelzenmuller, Julian Calliandro, Tristan Bettencourt, Andrea and Arthur Houkstra, and Nic Gihl.

Introduction

This is the cover photo from the first edition, the Sun appearing from behind Earth. The Apollo 12 crew took this photo in November 1969 on their way home from the Moon. Thank you to all those who made the first edition of this book a success. Many exciting things have happened in the past year, which I will describe in the Afterword. You will learn why 2015 is The Year of Light.

This new edition is dedicated to those who are traveling in discovery's footsteps. Starting with Chapter 2 and the Mediterranean, history is organised like an ocean voyage. You may follow the footsteps of the Greek thinkers in Athens or Galileo in Florence. You will hopefully see places that are off the beaten path, like Leonardo's village birthplace or Tycho Brahe's island observatory.

I hope you enjoy your voyage. As we move forward in Time, we will travel across the Earth and into the sea of Space. You will hear not just about light, but about gravity and whether our world is spherical.
All are connected by The Speed of Light.

1
A Babe in the Universe

A child's eyes open, and light floods in.

Our lives begin in darkness, the bubble of the womb. Within its confines a baby experiences warmth and pain without the light of understanding. At birth she emerges into a far larger Universe. With her first gasp a baby breathes an atmosphere transparent to light. Like a book opening, the Universe is revealed. Through light her eyes will see much of this world. Though it may seem cold and dark, we live in a Universe of light.

Each human birth is the culmination of many events. A baby has grown from a tiny cell barely a millionth of a meter across. Before her parents met, a parade of ancestors walked the Earth. Her human species resulted from evolution of life over billions of years. Before life could take hold on Earth, the Sun and Solar System formed in Space. A Milky Way galaxy formed with hundreds of millions of stars. The galaxy is part of an enormous Universe that itself grew from a tiny point. Like the tip of a cone, many events converge to create a single human life.

Astronomers celebrate "First Light" when a new telescope is first opened to the Universe. A baby experiences First Light when her eyes open. Images, sounds and language rapidly accumulate in her growing brain. Watching a child grow is witnessing the process of discovery.

Much of a baby's first months are spent staring, especially at faces. Infants are quite nearsighted, and can focus no more than 30 centimeters away. Almost immediately she can recognise faces that come into view. Humans are so well programmed to recognise faces that we imagine seeing them in the surface of the Moon or Mars. Discovery causes a baby to smile--an expression so universal that it need not be taught.

During a baby's first few months she learns causes and effects. She finds that crying can bring a response from others. She becomes conscious of her hands and feet, and begins to learn their uses. She learns about gravity by falling down many times. An instinct to explore drives her to crawl and eventually to walk.

From this curiosity comes a lifetime desire to learn. It would be impossible to learn all of knowledge from scratch, so humans build upon the lessons of others. Humans often tend to do things the way they have been done before, and repeat answers they have heard before.

In this manner knowledge is passed from one generation to another.

Language is another important skill, for the world seems less mysterious when things have names. Once a thing is named, she can attach other words to further describe it. She may someday learn mathematics to describe the world more precisely. These languages of words or numbers are tools for understanding.

She can be heard talking quietly in a language that only a baby understands. Even after learning from others, she remains capable of finding her own descriptions of things. The ability to find her own answers is a key to expanding human knowledge. Time and again, we will see that advances begin from the light of a single individual.

Around her second birthday a child has earned concepts of Space, can associate the kitchen with eating and her bedroom with sleeping. She still lives chiefly in the present, seeming to want everything right now! By her third birthday, she has learned about before and after, the beginnings of her understanding of Time. Much later humans learn to connect Space and Time, coordinating their actions to catch a ball or arrive at a place on time. We will see that Space and Time are linked by light.

Pictures in a book are an early source of interest. Her eyes unconsciously connect the two dimensions of a picture with the three dimensions of her world—the picture of an apple is an apple in her imagination. Pictures in books illuminate a new and magical world for her eyes to see. The writing beneath the pictures soon draws interest, too. Through reading a child can access the sum of human experience. Opening a book brings light to the world.

A child's mind is full of questions. The curious child will repeatedly ask her parents or teachers, "why?" An answer will just get another "why?" as the child seeks a deeper meaning. A new object or animal elicits a curious "What's that?" A mind that continues learning about the Universe also continues to grow.

When faced with something new, humans often find temporary answers. At some time a child may be told that babies are delivered by a stork. Later she learns the truth is far more exciting. Humans have often adopted temporary explanations, such as believing that Earth is centre of the Universe. Growth requires the ability to abandon a theory when a better one comes along. The light of science advances when theories are improved.

Much of human history was spent huddling in shelter at night with darkness all around. The only light would have been the Moon, stars and an occasional firefly. Creatures lurking in the night seemed mysterious and dangerous. Humans are instinctively drawn to light and knowledge. The rising Sun or the fires of home are sources of comfort and warmth. Scientists have spent centuries studying, arguing and wondering about light.

The process of discovery follows a pattern learned by babes. We are presented with a world that at first seems confusing. Gradually, we extend our senses and minds to understand the Universe. We learn to see patterns, connections, causes and effects. Often we learn from lessons of others before us. Sometimes we adopt a temporary explanation until a better one comes along.

When an individual comes up with an idea, the lesson may be passed on to others and all humanity advances. A baby's influence radiates outward like starlight. Eventually her life affects a whole community. As she grows into adulthood, a human has the potential to affect events across the planet and possibly beyond.

Humans share many of the same questions, starting with who they are and what is their place in the world. How big is the world and what is its shape? How did the world come to exist? How were the Sun, Moon and other sky objects created? How did the Universe begin? These questions are so universal that they have been asked across history.

A child's curiosity leads to a history of exploration. We will start with the discovery that Earth is curved and not flat. The question whether Earth was centre of the Universe is an important lesson for today's science. From there we move outward to discoveries about the very nature of Space and Time. The phenomenon of light is an intimate part of this Universe. We will see old questions about light itself, such as its speed or whether it forms waves or particles. The phenomenon of light is key to discoveries from Relativity to the tiny quantum world. We will see that our Universe began with a burst of light.

I had the privilege of discovering a change in the speed of light. From studies of the Moon at NASA, evidence was found that light slowed exactly as predicted. Some of the data was recently published in a refereed scientific journal, and will be described in Chapter 8. Light is key to understanding our Universe.

Our exploration of the Universe begins at First Light. The world we experience is the result of many events coming together. Just as a child grows, human knowledge also grows. Discoveries about this Universe come with the speed of light.

NEXT: The Light of Exploration

2

<u>The Light of Exploration</u>

Map of the world by Jan Stoopendaal, 1730

Life begins in the sphere of the womb, but once upon a time humans may have thought their world was flat. That was an understandable assumption when their "world" was very small. For most of human history, few people ventured further than a few kilometres from their place of birth. For planning a home, farm, or even a town we need not take into account Earth's curvature. Even if one lives on a plain or by the seashore, the horizon appears to vanish into the distance like an infinite flat

17

plane. For most of the humans who have lived, it was unnecessary to think of Earth's shape. Scientists have at times thought that the entire universe is flat, like the Earth

She began life as a curious child, then grew into a beautiful and intelligent woman. Despite the responsibilities of children and adulthood, she retained her curiosity about the world. Many days she spent staring at the sea, wondering what lay beyond. Having seen babies grow into adulthood, she wondered whether the world had a size and shape. Though the woman's name is lost to history, she made one of the greatest discoveries of all.

The theory that Earth is round may have been first made by a woman waiting for a ship to return. Peering anxiously at the horizon, she would have seen a ship's sails appear before the hull. While waiting for days, she may have contemplated the curve of her breasts, seeing how a point on her skin could disappear around their curve. If she were pregnant, she would have seen her belly grow from flat to round. Comparing her observations, she might have suspected that she lived on the surface of an immense sphere.

From the point-of-view of a sailor on the ship, the land that he has set out from sank below the horizon. The sight of one's home disappearing into the sea must have been quite alarming! The first sailors to witness this effect must have turned and hurried home. Upon the sailor's return, the woman would claim that it was the ship that appeared to sink. They had both disappeared relative to each other's frame of reference. Discovery of Earth's curvature began a quest by men and women that would someday lead to a Theory of Relativity.

Since the discovery of Earth's shape, many women have waited for their ship to come in. They have waited for exploration to bring food to the table or a better life for their children. They have waited for a time when their own horizons would be greater—when a woman's ideas could be heard. That ship may soon arrive.

The Mediterranean Sea

Sailing the Mediterranean, you are quickly out of sight of land and in open sea. To ancient mariners it seemed endless, the "Middle of the Earth". The Phoenicians are thought to be the first to navigate beyond the horizon. Movements of the Sun and Moon determine East and West. The star Polaris, around which other stars seem to rotate overnight, points the way North. Astronomy was also used to predict solstices as an aid to farming. Knowledge of the stars led to a better life on Earth.

Statue of Pythagoras on island of Samos

In school we hear about Pythagoras via a theorem about triangles, but he was also interested in spheres. Pythagoras was born around 580 B.C. on the island of Samos. Today on an excursion to Samos you can see his statue in a village called Pythagoreio, with a view of the

Turkish coast across the sea. Pythagoras is said to have travelled the world in search of knowledge, from Greece to Egypt and possibly India. The so-called Theorem of Pythagoras was known in Babylon and India centuries before Pythagoras lived; he may have learned it in his travels and spread it to the West.

Pythagoras encouraged men and women to have diverse interests, making contributions to music and astronomy. To please his musician's ear, Pythagoras sought a "cosmic harmony." As a musician, he is credited with the idea that "music of the spheres" described the planets. Reasoning that it was the most harmonious shape, he theorised that Earth was spherical. Pythagoras is even said to have thought that Earth revolves around a central fire, the Sun.

Pythagoras is said to have had a wife called Theano. She was an accomplished scholar in her own right, publishing works on a variety of subjects. Her work *On Piety* mentions that Pythagoras thought that all things come from numbers, meaning that everything in the world could be described by equations. This idea is the basis of modern physics. Pythagorean ideas began a quest that would last thousands of years, to find equations describing the Universe.

The Agora in Athens with Acropolis in background

Athens, Greece

On an excursion to Athens today, you are in the footsteps of the Greek philosophers. The Acropolis dominates the city as it has for thousands of years. The Parthenon and its surrounding museums remind us that Greece was once the centre of Western thought. The ancient Agora was once the city's centre of business, shopping, seeing others and being seen.

Two centuries after Pythagoras, the great thinker Socrates taught his followers to question what they were taught. Socrates, who was born in 470 B.C., lived his entire life in Athens. He had no school, but haunted the Agora constantly questioning the beliefs of others. In his wake he gained a following of young people who enjoyed seeing him challenge the privileged classes. Accused of corrupting the young, Socrates was forced to drink a poison hemlock.

Socrates left no writings of his own—all we know of him is from the works of his followers. Plato wrote that Socrates' prison was within walking distance of the Acropolis. Nearby in Philopappou Hill, also known as the Hill of Muses, are some cell-like caves carved from rock many centuries ago. A sign marks these caves as "Prison of Socrates". We are not sure if Socrates was indeed imprisoned here, for other locations have claimed to be his jail.

If you are not in Athens, there are other ways to follow in Socrates' footsteps. He taught us to question everything, even what is "commonly known". If today we hear that the universe is filled with epicycles, repulsive energies, or tiny strings; we should question what we hear. Socrates believed in principles enough to drink the poison hemlock.

Plato's Academy envisioned by Raphael.

One of the young followers of Socrates, Plato, founded his Academy in Athens just outside the city walls. You may stroll its paths in Plato's Academy Park at Monasteriou 140. Around the site are many stones among which Plato or his students could have sat. In the tree-shaded grounds, you can easily imagine Plato walking alongside discussing philosophy or science.

One of Plato's sayings has great import today, "We can easily forgive a child who is afraid of the dark; the real tragedy of life is when men are afraid of the light".

Plato's Academy admitted both men and women. Two women are known to have attended, Axiothea and

24

Lasthenia. Plato thought that the world could be deduced from simple principles. In search of those principles, he tried to relate the planets to geometric shapes. Plato studied the writings about Pythagoras, and taught the spherical Earth to his own students.

Plato's student Aristotle founded his own school, called the Lyceum. The ruins of the Lyceum were uncovered in 1996. You can see them near the intersection of Rigillis and Vasilissis Streets, close to the National Gardens. The Athens War Museum is also next door, so you can see Aristotle's Lyceum as a side-trip. The remains of the Gymnasium, where students trained for combat, are easily visible.

Aristotle compiled more evidence of a spherical Earth. The Southern constellations appear to rise higher as one travels farther South. The shadow of Earth upon the Moon during a lunar eclipse is circular. Aristotle was fascinated by symmetry and repetition. Teacher and disciple relied on different methods--while Plato preferred to deduce the world from first principles, Aristotle looked at physical evidence. As the teachings of Plato and Aristotle have spread through time, cosmologies based upon principles have sometimes conflicted with those assembled from measurements.

Aristotle believed that the heavens were filled with a

substance called quintessence, which was invisible and filled all of Space. He would not be the last to assume that the Universe is filled with an invisible substance. Even today scientists are tempted to explain the universe with quintessence. Aristotle also assumed a cosmology where Earth was the centre of all things. This flawed view of the Universe would last nearly 2000 years.

One of Aristotle's students was the restless son of King Phillip of Macedonia, who later would be known as Alexander the Great. His conquests would spread Greek learning throughout the known world. After his conquest of Egypt, the city of Alexandria was founded. The centre of Greek thought sailed East to Egypt.

The Library of Alexandria rises again

Alexandria, Egypt

In 2002 a modern Library of Alexandria was opened
on the shore of the Mediterranean. It is an excellent place
to visit and learn about the original. The Library of
Alexandria was founded around 295 B.C. by Demetrius
of Phalerum, a student of Aristotle. He convinced the
city's ruler that Alexandria could succeed Athens as a
centre of learning. Ships entering Alexandria 's harbour
saw their scrolls and maps seized and copied for the
Library.

Eventually the Library collected around 700,000
scrolls on everything from religion to geography. So
many scrolls were collected that a second repository was

opened at the temple of Serapis. More than just a collection of scrolls, it also served as a university, attracting scholars from around the Alexandrian world. In one place were Hebrew scriptures, Buddhist texts, Egyptian astronomy, and the complete plays of Sophocles and Euripides. Librarians at Alexandria had access to the amassed knowledge of the ancient world.

By 240 BC, many educated people believed that Earth was spherical. In that year Erastothenes, the Librarian of Alexandria, made a remarkable estimate of Earth's size. Making use of Pythagorean geometry, Eratosthenes combined the altitude of the Sun at different locations on Earth with estimates of the distance to those locations. He derived a circumference of 250,000 stadia. Though the exact length of a Greek "stadium" is not known today, Erastothenes' figure was amazingly accurate.

Aristarchus, a later librarian of Alexandria, published two books about the Universe. His work On the Sizes and Distances of the Sun and Moon contained estimates that were strikingly good for their time. Those distances were estimated from the Earth, fitting an Earth-centred cosmology. The enormous distances may have started Aristarchus thinking about alternatives. He wrote another book that is now lost and known only through other writers.

This second book introduced a cosmology with the Sun in the centre and Earth circling as a planet. Aristarchus also believed that the stars were immeasurably distant, so far that they appeared fixed in the sky. This second book caused great controversy. Another writer angrily suggested that Aristarchus should have been put on trial! The cosmology of Aristarchus was incredibly prescient for its time.

Hipparchus, another scholar at Alexandria, calculated the distance between Earth and the Moon. He also believed that Earth circled the Sun. The geographer Ptolemy also had the good fortune to study in Alexandria's library. Ptolemy published the greatest compendium yet of information about the spherical Earth, his *Geographia*. Nearly all of Ptolemy's work was copied from others, the scrolls and maps seized from passing ships. He also published the Almagest about the planets. The star positions in Ptolemy's *Almagest* were taken from Hipparchus.

Copying Aristotle, Ptolemy made Earth the centre of his cosmology. From Ptolemy until the time of Copernicus most people would believe that Sun, planets and stars circled the Earth. In some ways this was another understandable assumption. A navigator on Earth need not take into account the distances to stars, for they are too great to affect her calculations. Many astronomers, including this one, keep on the shelf a clear globe with stars depicted on its surface. Though we realise today that this model is just a convenience, for most of history humans believed that stars were truly fixed to an immense sphere.

Anyone observing the planets for a length of time will see them appear to reverse course and travel backwards in their paths. To explain this retrograde motion, Ptolemy's cosmology relied upon epicycles, spheres within spheres. The planets were each attached to an invisible sphere, in turn attached to bigger spheres surrounding the Earth. The epicycles supported Ptolemy's Universe, preventing it from crashing to Earth.

In the twilight of the Library, one of the most notable and lovely figures at Alexandria was Hypatia. She was daughter of Theon, one of the last scholars at Alexandria. Hypatia was renowned not just for her beauty, but also her accomplishments in mathematics and astronomy. She is said to have translated part of Ptolemy's *Almagest*. Hypatia made public appearances throughout the city, drawing crowds to her talks on science. Her increasing influence on Alexandria's governor made others jealous. Tragically Hypatia fell victim to a mob in 415 A.D.

A popular book about the *Cosmos* imagines Hypatia at

the Library's burning, vainly trying to stop the destruction of knowledge. In fact over its lifetime the Library suffered multiple fires, from the siege of Alexandria by Julius Caesar in 48 B.C. to conflict with Christians and Muslims in the 5th century. The original Library of Alexandria was gone by the time of Hypatia's death. The Temple of Serapis, which housed the last scrolls from the Library's collection, was destroyed in 391 A.D. Today in Alexandria you can visit a grand new Library near the site of the original.

The Ptolemaic cosmology was carefully assembled from observations. As knowledge of Earth composed the earthly portion of the *Geographia,* Ptolemy used the available observations to explain the wider Universe. The thoroughness of Ptolemy's assembled observations would have pleased Aristotle. Had the master Plato reviewed Ptolemy's work, he might have complained that this cosmology of many epicycles lacked an overriding principle. There was no single law to explain the motion of heavenly bodies, just epicycles. The world would see the rise and fall of a Roman Empire, the Dark Ages and Renaissance before Ptolemy's cosmology was overturned.

View of Torun from across river

Torun, Poland

The old town of Torun, about 100 miles from the port of Gdansk, looks very much like it did in the time of Copernicus. Torun is worth an excursion, for the streets and buildings are a picturesque and well-preserved. The medieval atmosphere is made safer because motorcars are not allowed on the streets. The tower of the Old Town Hall has an excellent view of Torun and the countryside. Torun even has its own Leaning Tower. You can easily imagine Copernicus in priestly garb hurrying by on business or watching the sky from church walls.

This writer was pleased to learn that Nicholas Copernicus was also born on February 19. At 15/17 Copernicus Place you can find the house owned by

Copernicus' family, thought to be his birthplace in 1473. Today it is a museum and a fine example of a Polish merchant's home. At the Cathedral of St. John the Baptist you can see the font where the infant Nicholas was baptized. A statue of Copernicus stands in front of the Old Town Hall and Nicholas Copernicus University. The University Observatory contains the third largest radio telescope in Europe, and a planetarium.

Many times we will see that the most unlikely of individuals can change the world. For those who knew him, the man who would start a revolutionise astronomy must have seemed quite un-revolutionary. Nicholas Copernicus moved through society in the honoured position of a priest. This quiet man of the church led a secret life of science—his nights were spent on church walls observing the sky. Quietly he collected and checked his observations for a book, but delayed publishing until the end.

De Revolutionibus Orbium Celestium, "The Revolution of the Heavenly Spheres" proposed that Earth is not the centre of everything, but circles the Sun. Possibly Copernicus delayed publishing for fear of ridicule— people would tell jokes about the Polish astronomer. Copernicus also felt his theory was incomplete, for the planetary orbits could not fit into perfect circles. The book was finally published in 1543, the day before Copernicus died quietly in bed.

Copernicus was buried in an unmarked grave beneath the floor of the cathedral in Frombork, where he had once served as canon. The town, also known as Frauenberg, is about one hour's drive from Gdansk. Only in the first decade of the 21st century were Copernicus' remains discovered and identified. In 2010 he was reburied with full honours. Today in the Frombork cathedral you can see his grave decorated with Copernicus' vision of the solar system.

(Copernicus has inspired more than one monument. A statue of Copernicus by Bertelsmann Thorvaldsen stands at Krakowskie Przedmiescie in Warsaw, in the square before the Polish Academy of Science. The new Copernicus Science Museum, which also houses a planetarium, is several blocks downhill on the banks of the Vistula. The original copy of Die Revolutionibus Orbium Celestium, signed by Copernicus on his deathbed, is kept at the Jagellonian University library in Kraków).

A quiet end was not in store for another man of the church. Giordano Bruno was born 5 years after Copernicus died, trained as a Dominican priest, and eagerly embraced Copernicus' new theory. Bruno's own ideas went even further—he speculated that we live in a boundless universe, that our solar system is just one of countless others, and that these other solar systems could be home to life! Thinking about other worlds, Giordano Bruno was far ahead of his time.

His seemingly heretical ideas caused Bruno big trouble with authorities. Unable to lead a steady life, Bruno taught and travelled from Italy to Switzerland to England. Tragically, a local official of the Inquisition ordered him executed. Today you can see Giordano Bruno's statue amidst the flower-sellers in Rome's Campo di Fiori, where he was burned at the stake. Revolutionary ideas nearly always meet with opposition.

ORTHOGRAPHIA PRÆCIPVÆ DOMVS ARCIS VRANIBVRGI
in Infula Porthoni Danici Venefia, Aufee Huenna, Affensaner veflaurande gratis, circa annum MDLXXX
à TICHONE BRAHE   ædificata

Uraniborg main building. Copper etching from Blaeu's Atlas Major, 1663.

Uraniborg

Copenhagen, Denmark

   In the port of Copenhagen you can see the Tivoli
Gardens and the many palaces built for Denmark's royal
family. Kronborg Castle, the Elsinore of Shakespeare's
Hamlet, guards the shore of the Oresund, the strait
between Denmark and Sweden. From the Round Tower,
an astronomical Observatory built in the 17th century,
you can see across the Oresund into Sweden.

   Copenhagen's most-visited statue is of Hans Christian
Andersen's *Little Mermaid*. Smaller ships may dock at
Toldbod quay, close to the city centre between the

36

Museum of Danish Resistance and The Little Mermaid. You may be reminded that Hans Christian Andersen also told us of The Emperor's New Clothes. The little boy in that story was ridiculed for pointing out the obvious. The Emperor's invisible garments, admired like epicycles by the kingdom's learned elite, didn't exist.

On the shore of a lake, the castle-like form of the Tycho Brahe Planetarium is dedicated to Denmark's most renowned astronomer. Tycho Brahe was born into Denmark's nobility, but his lifelong passion was astronomy. One night in 1572 he saw a new star appear, so brilliant that it could be seen in the daytime. Tycho's supernova was an exploding star, one of the most violent events in the Universe. In its brief moment of glory a supernova can outshine an entire galaxy. Just as some people thought that Earth or the speed of light was fixed, they once thought that the stars were forever fixed in a celestial sphere. This new star seemed to indicate that the sky was not immutable.

Denmark's King was so impressed by Tycho that he financed a state-of-the-art observatory for him on the island of Hven, in the Horesund between Zealand and Tycho's home province of Scania. Uraniborg was a castle dedicated to astronomy, surrounded by formal gardens which can be seen today. The observatory was equipped

37

with the finest astronomical instruments that money could buy, though the telescope had yet to be introduced. When the winds of Hven were found to disturb the instruments, Tycho built another observatory underground and called it Stjerneborg. Construction of Uraniborg was said to have consumed about one percent of Denmark's budget. (In comparison, NASA accounts for less than one half a percent of the U.S. budget).

Though the island of Hven and Scania are today part of Sweden, you can visit the remains of Uraniborg. The island can be reached by ferry from Copenhagen or from Landskrona in Sweden. The All Saints Church in Hven contains the Tycho Brahe Museum, with replicas of his astronomical instruments. The museum also maintains Tycho's underground observatory Stjerneborg. The grounds of Uraniborg and its formal gardens are slowly being restored. Today at the Cafe Tycho Brahe, adjacent to his observatory, you can dine on vegetables and herbs grown in Tycho's garden.

Florence, Italy

When you visit Florence, a 90-minute drive from the port of Livorno, you are in the centre of the Italian Renaissance. The Duomo of the Cathedral of Santa Maria del Fiore dominates the skyline as it did in the 16th century. The Ponte Vecchio was the first bridge across the Arno, and the only bridge in Florence to survive World War 2. Exploring the Uffizzi Museum, grandest art collection in Italy, you will see masterpieces by Titian, Canaletto, Leonardo Da Vinci and other residents of Florence. When you enter the Uffizzi courtyard from the Piazza Della Signoria, Galileo's statue is in the corner to your right.

Born in 1564, Galileo Galilei was fascinated by experimentation since childhood. Though he was an excellent student, financial reasons forced Galileo to leave university before graduating. He found work at the University in Padua, where you can still see the podium he taught from. While teaching in Padua, he preferred performing his own experiments to copying from books. Most likely he learned of the invention of the telescope from sailors, and was the first to turn it toward the sky.

On the bank of the River Arno, nestled alongside the Uffizzi Museum and in sight of the Ponte Vecchio, is the Museo Galileo. Far smaller than its neighbour, the museum can be experienced in 1-2 hours. After entering, you may head up the stairs to Room I, which is devoted to the Medici collection. The Medici family, rulers of Florence, acquired a huge assortment of astronomical and navigational instruments. Mastery of navigation allowed city-states to extend their influence.

Room II is devoted to astronomy and time. At the centre of the room, surrounded by navigational instruments, is a globe dating from 1085 A.D. Room III is dominated by an Armillary sphere. Exquisitely fashioned by master craftsmen, the many rings and circles track the movements of planets in the sky. With Earth at its centre, the Armillary represents Ptolemy's view of the universe. In the 14th century a translation of Ptolemy's *Geographia* appeared in Italy. Interest in Ptolemy included his Earth-centred cosmology.

When you enter Room VII, to your right is a case containing Galileo's original telescopes. Through his telescopes Galileo was first to see craters on the Moon, the phases of Venus and the Rings of Saturn. In the same case is the objective lens through which Galileo was first to behold the moons of Jupiter, circling something other than Earth. Though Galileo was initially skeptical of Copernican theory, observations led him to eagerly accept it.

Room VII also contains Galileo's fingers! A century after his death, some of his followers preserved his middle finger for posterity. Today Galileo's finger continually points upward, beckoning us to the sky. The display case also holds two other fingers and a tooth that were found later. In one room you can see not only Galileo's telescopes, but also the fingers that held them.

Galileo's observations were a threat to Ptolemy's system, because they showed objects circling something other than the Earth. As Plato had complained before, some men were afraid of the light. Learned minds of the day refused to peer into his telescope, for fear of upsetting their worldview. Galileo's beliefs led to his trial and sentencing for heresy. He was forced to publicly renounce his beliefs and spend the last decade of his life under house arrest.

The 14th century Church of Santa Croce, Italy's largest Franciscan Gothic Church, is a short walk from Museo Galileo. When you enter the nave from the ticket office and turn right, you can see Galileo's grave on the right side near the back door. Within the Church you will also find the graves of Michelangelo and Niccolo Macchiavelli. Persecuted in his time, Galileo is today recognised as one of these great Florentines.

Pisa, Italy

Pisa, once a rival city-state with Florence, is 15 miles from Livorno's port. An ambitious traveler on a day's excursion from Livorno can quickly pass through Pisa on the way to Florence. Galileo was born in Pisa, and grew up in his family home on Via Giuseppe Gusti. You can still see the Galileo family house, now marked by a plaque.

Until the 13th century Pisa was unequalled as a maritime power. Aided by a strategic location near the mouth of the River Arno, Pisan ships controlled the Mediterranean. In this period Pisa built the Campo de Miracoli, the Field of Miracles. On the lawn you can see the Baptistery, Campo Santo, and of course the Leaning Tower. From the Leaning Tower Galileo reportedly experimented with dropping different objects to see if they fell at the same rate.

Within the Pisa Duomo hangs a bronze fixture called "Galileo's Lamp". A lamp in the cathedral identical to this one supposedly inspired Galileo to think about pendulum motion. Galileo realized that the period of swinging was regular, no matter how far the lamp swung. The lamp started Galileo thinking that a pendulum could be the basis of an accurate clock. In his experiments, Galileo could have used a good clock.

Galileo also suggested a way to test the speed of light. For most of history humans did not know whether light had any speed at all. Aristotle believed that light simply traveled instantaneously. This is an understandable assumption, for light travels so quickly that to observers on Earth its speed seems infinite. Galileo suggested stationing observers with lanterns on distant hilltops to time light's passage, but he still lacked a good clock. A careful scientist, Galileo could only conclude that if light had a speed it was too great for him to measure.

Though today we agree that Earth is not the centre of the Universe, the Copernican model was not complete. The orbits of planets are not perfect circles. The

Ptolemaic cosmology could explain this by imagining more epicycles. Scholars kept themselves employed inventing theories with up to 100 epicycles. The mathematics was highly complex, which kept discussion limited to an educated few. Detractors could point to the non-circular motion of planets and say that Copernicus was wrong. Copernicus even tried to put epicycles in his own theory. Without a guiding principle, the Copernican system was open to criticism.

For most scholars, it was far easier to promote an old system. In addition to curiosity, scientists have the same desires as most other people. One of those desires is to belong to a larger community. Another desire is security, the ability to keep a job and pay the bills. To maintain place and security, it is always easier to support the fashionable theory.

A secure life was not in store for Galileo's friend Johannes Kepler. Though today we know him as a brilliant mathematician, in his lifetime Kepler moved from job to job before becoming Tycho Brahe's last assistant. Tycho's treasure of observations was kept secret from everyone, even his assistant. Only after Tycho's death did Kepler gain access to all the data. Kepler believed in the Copernican system, and sought to find a principle behind it. Like Pythagoras he thought that mathematical relations underlie all of nature.

Fascinated by Plato's geometric forms, Kepler tried to explain orbits as ratios of these shapes. He had greater success with ellipses. The Indian scientist Aryabhatta, who was born in 476, had independently suggested that planetary orbits are elliptical. Starting from the principle of planets orbiting the Sun, Kepler concluded that the orbits are ellipses with the Sun as one focus. Kepler devised laws that described the motion of planets far better than any number of epicycles. Kepler's laws allowed him to predict the transit of Venus in 1631.

Planets in their elliptical tracks are like roller coasters circling the Sun. At the farthest point from the Sun they slow like cars atop a hill. Falling closer to the Sun they gain speed like cars coasting downhill. Nearer the Sun objects orbit faster and farther from the Sun their velocity slows.

After Kepler, the revolution was not complete. There was still no mechanism to explain what held the planets in their elliptical courses. This guiding principle would come from a Law of Gravitation.

NEXT: Moons, Apples and Light

3

<u>Moons, Apples and Light</u>

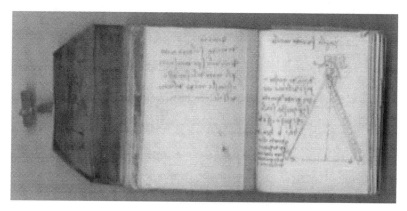

Leonardo Da Vinci, *Man Climbing a Ladder*, c. 1494, displayed at Victoria and Albert Museum 2006-2007.

Vinci, Italy

The hilltop Tuscan town of Vinci, about 25 miles from Florence, named its most famous son Leonardo. The town's 13th century Castello dei Conti Guidi houses part of the Museo Leonardiano, with replicas of Leonardo's inventions. Behind the castle is a wooden sculpture based on Leonardo's Vitruvian Man. The Piazza Dei Guidi, filled with sculptures inspired by Leonardo's science. In the Piazza Della Liberta is a bronze horse base upon Leonardo's designs. Sculptures in the Church of Santa Croce celebrate his baptism.

Leonardo Da Vinci birthplace

Leonardo was born 2 miles from Vinci in the tiny village of Anchiano. You can walk there via the Strata Verde, the Green Road that that Leonardo walked. The modest stone house where he was born is now a museum. The first room you see contains a very old fireplace, a wooden table and a bust of Leonardo. To your left is a darkened room where Leonardo himself greets you. For a moment he seems as real as an actor, but this Da Vinci is a life-size hologram. As an artist with a lifelong interest in light, Leonardo would have enjoyed this modern illusion.

The term "Renaissance Man" or woman came to describe someone accomplished in multiple fields. Pythagoras encouraged his students to be interested in everything from astronomy to music. Men and women who make breakthroughs tend to think out of the box by having many interests. A child is naturally curious about everything—it is hard to keep her mind focused on one thing! People working outside their fields—such as astronomers contributing to geology or vice versa, have made many contributions to science.

Leonardo Da Vinci was a true Renaissance man, accomplished as an artist, scientist, architect and inventor. As a master painter, he had a lifelong fascination with light and colour. His detailed drawings of human anatomy and fluid flow were worthy of a modern textbook. His drawings are a testament to Leonardo's inventiveness and scientific curiosity.

One Leonardo drawing is a study of light rays reflected from a parabolic mirror. On the same page is a tubelike device on a moveable mount, the exterior of a telescope. Among Leonardo's studies of military hardware is a drawing of cannonballs following parabolic arcs, predating another of Newton's studies. Leonardo had trained as an artist since childhood, and lacked the training to put his thoughts into numbers.

A century before Isaac Newton was born Leonardo was fascinated by gravity. He tried to design a perpetual motion machine driven by gravity, blocks and tackles for lifting, parachutes and flying machines for defying gravity.

His drawing of a man climbing up a ladder has a vertical line through the man's centre of gravity representing downward force. If this line had been extended to Earth's centre, and Leonardo had known more math, it might have led to a theory of gravity.

Gravity holds our Universe together, yet we are still learning its secrets. Our knowledge of gravity has advanced in leaps each preceded by an unsolved mystery. Isaac Newton's gravitation solved the mystery of planetary motion, and is still guides our spacecraft. Newton's discoveries have implications in the 21st century and beyond.

Newton home in Woolsthorpe. Note window in upper right.

Isaac Newton was born in 1642, the year Galileo died under house arrest. Though he was initially considered a poor student, Newton showed an aptitude for science at an early age. Like Leonardo he became fascinated by machines, and built a water clock when clocks were still rare. His uncle had studied at Cambridge, and convinced his family to send the young wizard to Trinity College. The distance from Woolsthorpe to Cambridge is about 70 miles, an easy drive today but a long journey for a boy in 1661. The ancient halls, gargoyles and stone turrets of the university must have seemed magical to young Isaac.

When dining in the Great Hall, Isaac Newton sat at the bottom of the table as a subsizar, the lowest ranking. For his first three years Isaac Newton paid his way through university with menial jobs.

Newton would later write: "If I have seen farther, it is by standing on the shoulders of giants." His education was steeped in the ancient masters. During Newton's time Cambridge used many teachings of Aristotle, and the Copernican model was still gaining acceptance. In 1664 Newton was finally elected a scholar, but that same year the plague closed down Cambridge. He traveled back to Woolsthotpe, and in two years from 1665 to 1667 started his greatest breakthroughs in gravitation and optics. From humble birth, the plague and solitude, young Isaac Newton produced some of history's greatest discoveries.

Nearly everyone has looked at the Moon, and many have seen apples fall from trees. It took the mind of Newton to see Moons and apples guided by one universal law of gravitation. Earth and the planets orbit the Sun in Keplerian ellipses. A falling apple is also in an orbit, one interrupted by the Earth. If a hole were dug to allow its passage, the apple would fall like Alice through Earth's interior and out the other side before falling back and repeating that orbit.

The cover of Newton's book *Principia* is a diagramme of cannonballs being fired at various velocities, uncannily like Leonardo's earlier drawing. Newton imagined the cannonball traveling even faster, so fast that it followed a circular path around the Earth without falling back. Three centuries before Sputnik, Newton imagined an artificial satellite. A satellite in a higher orbit would have a smaller velocity. If an object reached an escape velocity, it could leave the Earth completely. So powerful is Newton's Law of gravitation that it still guides spacecraft to the planets.

Where Leonardo had drawn the gravitational force as an arrow leading downward, Newton extended the arrow to Earth's centre. Newton deduced that the gravitational force was proportional to the mass of one object multiplied by mass of the other. Along with his competitor Robert Hooke, Newton knew that this force was proportional to the inverse-square of the object's distance. To give an answer with the dimensions of force, Newton introduced his gravitational constant G. Newton's Law of Gravitation showed mathematically how the planets move in ellipses. This mechanism completed the Copernican cosmology. Though Aristotle held sway when Newton entered Cambridge, Newton's gravity was the final victory of Copernicus.

Following the publishing of *Principia*, Newton's friend and priest Richard Bentley asked a simple but profound question: If gravity attracts every bit of matter to every other, why doesn't the Universe collapse? Newton, despite his genius, could not answer this question. Much later this same puzzle would lead to Albert Einstein's greatest blunder.

During the same period when plague kept him from Cambridge, Newton also contributed to the study of optics and light. By observing light pass through a prism, he deduced that white light is made up of many colours. This answered a child's question of why the sky is blue. Shorter, blue wavelengths of light are scattered by Earth's atmosphere. Scattering of blue light makes the daytime sky appear blue. Though Newton believed that light is composed of particles that he called corpuscles, diffraction showed him that it also behaved like waves. Argument over wave vs. particle would rage for centuries. Redshift of light spectra would someday signal expansion of the Universe.

On a trip to Woolsthorpe you can see not just Newton's birthplace, but also the window through which light entered his thoughts. Through the upper right window in the photo, Newton studied the refraction of light. You can easily imagine Newton holding a prism in this light to see it refracted into colours. You can also see an apple tree like the one that supposedly inspired Newton. The University in Cambridge belatedly planted an apple tree to honour Newton, but his greatest discoveries belong to his humble home of Woolsthorpe.

Le Observatoire

Paris, France

Light's speed was first measured from the City of Light. The Paris Observatory is at 61 Avenue de l'Observatoire, a short distance from the Latin Quarter. The nearest Metro station is at Denfert-Rochereau, but a more scenic walk begins at Jardin du Luxembourg. From the beautiful gardens you may walk South along the Avenue past the Fontaine de l'Observatoire. Atop this fountain is a statue of four women holding aloft a bronze globe.

Our Sun King Louis XIV financed the Observatory, wishing to stay ahead of rival nations. On the floor in the Observatory you can see a line marking Zero degrees longitude. Britain's Royal Greenwich Observatory has a more famous line also marking the Prime Meridian. Location of Zero degrees longitude is completely arbitrary, so maps drawn in different nations would place it in their country. Ptolemy's *Geographia* placed it in the island of Rhodes. The United States measured longitude from Washington's Naval Observatory. As France and Britain competed to build empires, they also competed for the honour of having Zero in their territory. Eventually the influence of Britain's sea power led to Greenwich being adopted by other nations, though the French vehemently disagreed. Observatories and their uses for navigation were very important for competing nations.

Galileo had tried to measure the speed of light with lanterns on a distant hilltop, but lacked a good clock or a distant hill. The first evidence that light had a speed came from another Galileo discovery, the moons of Jupiter. When astronomer Ole Roemer arrived at the

Observatory in 1672, there was an anomaly in observations of Io. The moon appeared from behind Jupiter late. Roemer realized that if Io's moonlight had a finite speed, it would delay the moon's appearance. Using data from this observatory Roemer was first to measure the speed of light, though his estimate of 210,000 kilometres per second was 30% too slow.

Having made a great discovery, Roemer was unable to convince his elders. His boss Domenico Cassini was a distinguished astronomer in his own right, whose name would be given to a spacecraft orbiting Saturn, but he believed that light traveled instantaneously. After years of work, in 1675 Roemer was bold enough to present the results on his own. He also predicted that on November 9, 1676 Io would appear at 5:35:45 rather than 5:25:45 as astronomers had calculated. On that date Io emerged from behind Jupiter 10 minutes late, precisely as Roemer had predicted. Though Roemer was correct, Cassini and others insisted that there was no speed of light.

50 years would pass before other experiments verified that light had a finite speed. James Bradley in 1728 used stellar aberration, when the position of stars appears to change because of Earth's orbit around the Sun. French physicists Hippolyte Fizeau in 1849 and Leon Foucault in 1862 used a system of rotating disks placed on hilltops

miles apart to measure the speed of light. Since Roemer's time, many other experiments have been devised to accurately measure c. Why light travels at 299,792.5 kilometers per second, not faster or slower, remained a mystery.

Isaac Newton could contemplate the sky, imagine cannonballs traveling at orbital speed, and calculate that a satellite in a circular orbit would have a certain velocity at a given altitude. Roemer visited Newton in 1679; very likely they discussed the speed of light. Since Newton knew that Moons and apples are both guided by one law, and believed that light is made of particles, might he have suspected that light is also affected by gravity?

One consequence of the speed of light has been Black Holes, objects so dense that not even light can escape. In 1783, the year the Montgolfier Brothers launched the first manned balloon, a clergyman named John Michell deduced their existence from Newton's Laws. By equating the speed of light with Newton's escape velocity, Michell concluded that a heavy enough object would trap even light. Mathematician Pierre Laplace promoted the same idea in a book published in 1796. Interested in many things, Laplace also suggested that the Solar System condensed from a cloud of gas. Of all denizens of the

Universe, Black Holes are among the most mysterious and powerful.

Thomas Young was another polymath, an individual with many interests. Among his many achievements, he translated much of Egypt's Rosetta Stone into English. As a physician, he studied the human eye and concluded that it processes three colours of red, blue and yellow. The eye's structure may have given Young ideas about light itself. In a talk delivered in 1801, Young argued that light takes the form of waves. This appeared to contradict Newton, who favoured a particle theory of light. Since Newton's legacy was revered, Young's wave theory of light was not accepted for decades.

Maxwell, Light and the Hypothetical Ether

For most of history communication was limited by the speed of human travel. Starting in the 1840's the electric telegraph was introduced. During the 1870's Alexander Graham Bell would perfect the telephone, allowing human voices to travel in electric wires. During the 19th century communications began moving at the speed of light.

Much of our familiarity with light comes from the work of James Clerk Maxwell. The scion of a wealthy Scottish family, Maxwell's background could hardly be more different from Newton's. Like Newton, Maxwell started his best-known work while an undergraduate at Trinity College. His other interests included the Rings of Saturn and the kinetic theory of gases. Fascinated by light, he also made the first permanent colour photo! Maxwell built upon the experiments of Michael Faraday with magnetic fields, and the discoveries of Andre-Marie Ampere about electric current. Maxwell showed that electric and magnetic fields were both part of one phenomenon, electromagnetism. Electromagnetic waves can take the form of infrared, microwave, radio, UV, X-rays, gamma rays or visible light. Combining Faraday and Ampere's Laws with some equations of his own, Maxwell produced a set of equations describing all electromagnetic interactions.

Maxwell's equations allowed calculation of the speed of these waves. That speed of 310,000 km/sec was almost exactly the 19th century value of the speed of light. Maxwell thought this too close for coincidence. He proposed that light was also a form of electromagnetic waves. The same Maxwell equations also described the propagation of light! The work of Faraday, Ampere and Maxwell led to discovery that light can indeed take the form of waves.

Our lives begin in a tiny womb with no visible light—but we are not completely isolated from light's influence. The warmth we feel from our mothers is infrared radiation, another form of electromagnetism. When we feel heat, we feel a form of light stretched into long wavelengths. Certain animals, such as snakes, have eyes that see in the infrared. When we are drawn to warmth of a fire or another person, it is another way of seeking light

There has long been speculation of whether the speed of light has always been the same. In 1874, while Alexander Graham Bell was working on his telephone, William Thomson claimed that the speed of light was slowing. Thomson was one of the most prominent scientists of the late 18th century, honoured as First Lord Kelvin and via the Kelvin temperature scale. In 1874, with William Tait, Thomson published a paper claiming that light slows down. Though the subject is not often discussed, many experimenters have sought a change in c.

After Maxwell, a number of mysteries still remained about light. The debate remained whether it was particle or wave. Newton had believed that light was made of particles called corpuscles, even though his own experiments showed that it had wavelike characteristics. Maxwell's discovery that light is an electromagnetic wave appeared to settle the argument. The advent of quantum mechanics would show that they were both right.

Wavelike properties led to one of the biggest misconceptions about light. Since sound waves travel through air and ocean waves through water, it seemed logical that light also propagates through some sort of medium. This imaginary substance was called ether. Since light travels throughout the Universe, this ether was assumed to fill all of Space. Earth was believed to travel through the ether like a ship through water. Nearly every scientist of the 19th century, including Maxwell himself, believed in the ether.

Ether shared with epicycles all the hallmarks of a scientific misconception. Like Aristotle's quintessence, ether was presumed to be everywhere in Space. Since we can't see ether in the blackness of Space, it was assumed to be invisible, just like the epicycles. Ether's existence was only inferred--no experimenter claimed to have a test tube of it or even a microscope slide. The greatest of theorists could only speculate about the nature of ether, so that was left for some future experiment. Something that dominates Space, is invisible, but can't be isolated shows signs of another misconception.

The concept of ether was dealt a mortal blow by the experiments of Albert Michelson and Edward Morley. Both men were fascinated by the speed of light. If light traveled through the ether at fixed speed and Earth moved through the ether, than the speeds of light parallel and perpendicular to Earth's motion should have been different. Michelson and Morley's experiment involved an interferometer with two perpendicular arms. The differing speeds of light down the two arms would have created interference patterns when combined. To everyone's surprise, the experiment detected no such interference. The speed of light appeared the same in all directions. Even after multiple trials, Michelson and Morley failed to find any indication of Earth's motion through ether.

In the latter part of the 19th century, may theorists attempted to explain this result. Hendrik Lorentz introduced the mathematics that would lead to a solution. He also presumed that the speed of light through ether was constant. His Lorentz transformation was a set of mathematical equations for length and time. They suggested that lengths of objects are contracted in the direction of their velocity. In Lorentz's interpretation the Michelson--Morley apparatus literally contracted, making the speed of light appear the same in all directions.

As Newton's theory of gravitation solved the problem of planetary motions, his studies of light pointed toward other puzzles. Maxwell's equations showed that light was a form of electromagnetic radiation with a velocity c. The assumption that light waves traveled through ether, was contradicted by Michelson and Morley. Lorentz's theory that objects contracted with velocity was a huge step toward a solution. The principle that would explain this puzzle would not come from any professor or haughty doctor of science. It would come from a clerk in the Swiss patent office.

Next: If You Are Within the Sound of My Voice

4

<u>If You Are Within...</u>

Statue of Albert Einstein, Washington DC.

Today Albert Einstein's statue occupies a place of honour in front of the National Academy of Sciences, not far from the National Mall. Albert Einstein was another unlikely suspect to start a revolution. Like Newton, he would at times be considered a poor learner. When he was two years old, while Michelson and Morley were first performing their experiment, Albert had not learned to speak at all. The very act of putting his thoughts into words was a challenge for him. Growing up in a world of Prussian militarism, Albert grew to dislike the regimen of school. In his studies he preferred marching to his own drummer.

Like Newton and Da Vinci, Albert developed a love for gadgets. The workings of machines fascinated him. Possibly he inherited this fascination from his father, who ran an electronics business with Albert's uncle. When Albert was 5, sickly and still beginning to speak, his father gave him a magnetic compass. The movements of the compass needle in response to unseen forces began a lifelong fascination. In later years Einstein would call Earth's magnetic field one of the greatest unsolved problems of science.

Though Albert disliked competitive games, he led a rich life outside of school. Their dynamos and electrical devices built by his father were sparking, whirring magic to a boy. His Uncle gave him a book on algebra while Albert was still in elementary school. When Albert was 12 his uncle Max Talmud gave him a book on plane geometry. Einstein loved his "sacred little geometry book" as a child would a treasured plaything. The simple yet elegant proofs of Euclid and Pythagoras pleased a young mind seeking order. Einstein was inspired by

Pythagoras' search for harmony. His uncle Max also gave him Kant's *Critique of Pure Reason* when Albert was 13. When his school finally taught him algebra, Einstein was at least 2 years ahead of the class.

When Albert was 14 the business run by his father and uncle fell on hard times and was forced to close. The Einstein family moved to Milan in search of other opportunities. Albert had grown up with this business, and the loss affected him deeply. Later his science career would take him from Milan to Bern and eventually Princeton, uprooted from a sense of home. He learned to think independently, insulating himself from the fashionable ideas of his time. His thoughts and beliefs traveled with him, eventually reaching out to the Universe.

As a university student, Einstein preferred his own studies to the assigned work. He had a reputation for borrowing the textbook on the night before the exam to study. In his last year he ran afoul of the professor of experimental physics. Einstein's original suggestion of a way to test Michelson-Morley did not impress the professor, whose name is forgotten.

In 1900, academic success depended on pleasing a large number of the old and inflexible. The corollary is that the disfavour of one can cripple an academic career.

Though his fellow students all received recommendations for further positions, Einstein did not. A long period of unemployment led to a great deal of anxiety. The century's greatest scientist could not find a job in science.

With the help of family and friends, Einstein finally found a position at the Swiss Patent Office. Examining patent applications proved to be satisfying work for someone who had grown up with electronics. In this period, separated from the usual circle of academic discourse, Einstein produced his four great papers of 1905. Any one of those papers would be considered a huge breakthrough in physics. Through his trials, Einstein retained the sense of wonder that children share. He later wrote, "All physical theories, their mathematical expressions notwithstanding, ought to lend themselves to so simple a description that even a child could understand them."

Bern's best-known landmark is the Zytglogge, a clock tower that was once the city's Western gateway. The colourful astronomical clock was built in the 16th century. At four minutes before each hour the procession of mechanized characters begins. From 1903 to 1905 Einstein lived up the street at Kramgasse 49, and saw this parade many times. Einsteinhaus is now a museum where you can see his writing desk and other mementoes of time.

The turn of the 20th century was another exciting time for physics, the beginnings of a quantum revolution. This had begun with another mystery of light. A heated body produces radiation in a characteristic spectrum, like a hot piece of iron glowing red. At very low wavelengths this radiation seemed to vanish. This "ultraviolet catastrophe" defied Maxwell's equations. Max Planck spent years trying to solve this mystery. Finally in 1900 he solved the catastrophe with an "act of desperation," introducing the quantum value h. Multiplying by the speed of light c, the energy of light is quantized into tiny amounts by hc.

As a conservative man of science, Planck tried everything possible to avoid introducing h. After 6 years of work he could find no other way to explain the data. The Planck formula for blackbody radiation fit the data curve precisely. When a prediction line matches the data, it probably means something.

Quantum mechanics predicts that the tiniest particles cannot be located precisely. Such measurements are subject to the Uncertainty Principle. While we cannot locate a particle precisely, we can predict the probability that a particle will be in a given location. With Quantum mechanics, predicting the behaviour of particles becomes a game of statistics.

Surprisingly, Quantum mechanics predicts that Space itself can be filled with "virtual particles." Paul Dirac used QM to predict the existence of *antimatter,* a negative counterpart to matter. Statistics predict that under certain conditions, pairs of matter and antimatter particles can appear in a process called "pair production." Quantum mechanics helps explain how the matter that we are made of formed.

Einstein's first 1905 paper concerned the photoelectric effect. Light shining upon metal produces electricity, an effect that powers solar cells today. The behaviour of the power produced could not be explained by Maxwell, or by wave theories of light. Einstein showed that this effect could be explained if light were quantized by hc into particles called photons. This returned to Newton's belief that light takes the form of particles.

New ideas often have difficulty even getting published. In Einstein's time there was no "peer review," papers only had to be approved by an editor. Fortunately for the history of science, the editor of *Annalen der Physik* was Max Planck. He quickly recognised the importance of Einstein's idea and approved their publication. If not for Planck's interest, Einstein's papers might have been delayed indefinitely. After their publication, the reaction was still a deafening silence. Though Einstein was fortunate that his 1905 papers were promptly published, not until 1911 did the magazine *Scientific American* mention his work. Einstein would wait years before the wider world recognized his achievement.

Something in Albert's upbringing made him think different. He grew up late in speaking with others. He gained curiosity and love of science from his father and uncles. Somewhere in his journeys, he came upon a truth that children can learn and adults can forget. Space and Time are one phenomenon.

Humans are born hardwired to think this. In Galileo or Newton's time the distance between towns would be given in days traveled rather than miles. Today when you ask most people the distance from their work to home, very few know the distance in kilometres or miles. They nearly always know the time between those points, often to the exact number of minutes. In an age of clocks, this shorthand keeps us from being late to school.

The conversion factor between distance and time is the speed of our vehicles. When we are hurrying to work, our speed is usually the speed limit of the road. Nature has her own speed limit, built into everything from the tiniest atoms to the very Universe. That speed is commonly known as the speed of light. This value c is not just the speed of light; it is the speed of any massless particle and the speed of information. It was originally named for the Latin *celeritas*, but more accurately it is a *conversion* factor between Space/Time. Whether conversion is expressed in miles per hour or meters per second, it is a velocity.

The start of the twentieth century also saw the introduction of radio. The young inventor Marconi had demonstrated practical radios in 1897. By the end of 1901 Marconi had sent radio signals across the Atlantic. Nearly all of us have spent hours listening to a radio. Hearing music or news over the airwaves is another magical experience. The electromagnetic waves of radios, telephones and televisions respectfully obey Maxwell's equations. Disk jockeys often begin an announcement: "If you are within the sound of my voice…" In this phrase lay the truth that Einstein would struggle to put into words: Space and Time are one phenomenon related by c.

This can be expressed in this 2-dimensional page if the three spatial dimensions x, y, and z are compressed into one dimension, r. Now we can put the additional dimension of time into the vertical. At the centre is a radio station, broadcasting a signal at the speed of light. Seen in this drawing, the spreading of these electromagnetic waves appears as a cone, a light cone.

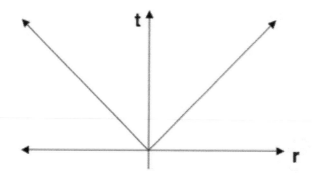

Events elsewhere in Space/Time can be either outside or inside this cone. An event on the Moon less than one second after our signal transmits is outside the cone. Since light takes more than one second to reach the Moon, our radio signal does not reach this event in time. The separation between these two events is said to be *Spacelike*.

If one second after our signal is transmitted, another rocket is launched from a point on Earth, that event is inside the light cone. Since it takes light less than a second to cross the Earth, our signal reaches this nearer rocket before it launches. The separation between these events is called *Timelike*. Their separation is literally a matter of time.

If the centre is a baby's birth, the cone represents her family broadcasting and tweeting the happy event. The expanding light cone represents the events that the baby's

75

birth will influence. The events that influence a baby's birth are also contained in a cone, converging at the birth. The event is at the apex of both a converging and an expanding cone of light.

Mathematician Johann Lambert in the 18th century found formulae for areas of triangles in curved spaces. He found that when applying those formulae to spherical spaces, the radius of those spheres was the square root of a negative number. These so-called imaginary numbers would turn out to be critical in Space/Time. The Timelike separation of an event within the light cone also turns out to be the square root of a negative number. For a spherical Space, mathematics favours a fourth dimension of Time.

To put Space and Time on the same page, we need a conversion factor. That factor shows us how much vertical Time to draw for each unit of horizontal Space. The conversion is provided by c. The value c is much more than just the speed of light, for the rules of Relativity apply even in a dark room with the lights out. Even when no one can see you, you can't exceed the speed of light.

From this principle, called *causality*, one can derive all the equations of Special Relativity. Mathematician E.C. Zeeman in 1964 showed that the causality principle led to the Lorentz equations of Relativity. The causality principle does not require the speed of light to be constant. Albert Einstein would later experiment with a varying c.

Seen in our 3 dimensions of Space and one of Time, the light cone looks like the famous picture of a radio transmitter. Marconi's experiments showed that an antenna works best when mounted on a vertical tower. Signals travel outward at the speed of light c. The expanding spherical waves of the radio transmitter are the light cone in four dimensions. Outside the spherical wave of a signal, a listener has Spacelike separation. Inside the expanding sphere, the separation is Timelike, within the sound of the announcer's voice.

Michelson and Morley's experiment appeared to show the same speed of light in every direction. Lorentz found mathematical equations to account for this, but could not find a principle behind them. Einstein added a simple stipulation: The speed of light is constant regardless of location. Einstein himself did not consider this to be a theory about light. Later he would write that a constant c was "neither a supposition nor a hypothesis about the physical nature of light but a stipulation which I can make at my own free discretion."

The predictions of special Relativity would fill physics books for a century. As Lorentz had first suspected,

77

objects traveling at high velocity literally contracted in length. The very passage of time would slow relative to stationary objects. A starship accelerated to near-light velocities could reach another solar system in a short time by the ship's clocks. Upon its return the crew would find that decades or centuries had passed back home. Since Einstein's time, many experiments have verified these predictions. As spaceflight became a reality, more of these effects of Relativity would be detected.

Relativity was not yet complete, as Einstein himself realised. It made no allowance whatsoever for gravity. Einstein would consider this problem for 10 more years. For several more years he would retain his patent office job, finally moving to a teaching position in 1909. Around 1907 he had a vision similar to Newton's, and of equal importance. "For an observer falling freely from the roof of a house there exists at least in his immediate surroundings no gravitational field," he would write. This was Einstein's Principle of Equivalence. He realized that an object in freefall feels weightless, an effect we later witnessed in orbiting astronauts. Einstein called this the most fortunate though of his life.

In 1911 Einstein spent a lot of time on a theory with a varying speed of light. By this theory, c would change in the presence of gravitational fields, causing the paths of

light rays to bend. Some people claim that Special
Relativity required c to be fixed, but Einstein himself was
willing to consider that it changed. Though his Special
Relativity stipulated that light's speed is the same
regardless of location, Einstein did not forbid it to change
over time.

Inspiration in physics requires the perspiration of
mathematics. Einstein was by his own admission not the
best mathematician, so he sought help from friends.
Marcel Grossman helped him with Riemannian
geometry. Contrary to the geometry that Einstein grew
up with, this was non-Euclidean. Riemann's vectors and
tensors allowed for spaces that were curved rather than
flat. Using Riemannian geometry, Einstein in 1915
published his General Theory of Relativity.

General Relativity stated that Space/Time is not flat,
but is curved by the presence of mass. Any massive object
causes Space/Time to be warped, as a stone placed on a
rubber sheet causes the sheet to dimple. A marble rolling
at high speed past this gravity well will see its course
slightly altered. If the marble travels too close and too
slow, it will fall into the stone's well. Given the right
velocity, the marble will circle around the well as a
satellite. Circling at a greater distance from the stone
requires a lower velocity. Gravity was not a force, but the

very curvature of Space/Time. Since light travels through Space/Time, General Relativity predicted that light itself would be bent by gravity.

Einstein's General Relativity solved a longstanding problem concerning Mercury's orbit. The innermost planet orbits in a Keplerian ellipse with an axis that precesses at 5,600 arc seconds per century. Many other objects exert small influences upon this orbit, but when all the influences were added there was a discrepancy of 43 arc seconds per century. In the latter part of the 19th century scientists tried many theories to account for the discrepancy. Again they hypothesized invisible actors--at one point an unknown planet called Vulcan was inferred to exist within Mercury's orbit. The curved Space/Time of General Relativity explained this mystery without imaginary forces. When his prediction matched the data Einstein was giddy with joy.

Special Relativity used the equations that Lorentz had already introduced, and may be considered as an idea whose time had come. In contrast General Relativity was a huge step forward, a revolution in our understanding of gravity comparable to Newton. Einstein did not set out to prove Newton wrong, rather he ensured that his equations reduced to Newtonian dynamics in general cases. By going beyond Newton, General Relativity allowed Einstein to envision the very structure of Space/Time.

Though Special and General Relativity are seen as two parts, they were never truly linked. Special Relativity still made no allowance for gravity. Reducing the equations of General Relativity to those of Special Relativity proved to be a daunting challenge. To link the two parts, one had to picture the shape that gravity would bend Space/Time into. Einstein dared to imagine the entire Universe.

Next: Einstein's Sphere of Light

5

<u>Einstein's Sphere of Light</u>

A book that everyone should read was titled *Relativity: The Special and General Theory* by Albert Einstein. The 1961 edition is subtitled *A Simple Explanation That Anyone Can Understand*. The book is available in paperback for less than 10 dollars, far less than overpriced textbooks. In his simple explanation Einstein uses concepts, like imaginary time, that are not taught in schools today. The shortcuts used in textbooks written by others hide the true beauty of

Einstein's Relativity. No one else understood the subject like Einstein himself.

In Chapter 31 Einstein attempts to imagine the entire Universe. Like Pythagoras 2500 years before, Einstein is motivated by a search for harmony. He follows what would be called the Cosmological Principle: By this principle the Universe looks the same no matter what direction one looks, and every bit resembles every other bit. He rejects a flat Universe, for his General Relativity shows that Space/Time is curved. A rebellious soul, he rejects boundaries and considers the Universe "finite yet unbounded". The obvious analogy is a sphere.

An ant on the surface of a sphere lives in a Universe that is finite yet unbounded. If the ant set off in a line of any direction she would eventually end up at the point where she started. From the ant's perspective, a large enough sphere would seem flat. That is an understandable assumption, for an ant's world is small. The spherical Universe is also closed yet without boundaries.

Einstein's General Relativity predicted that mass causes Space to be curved. If the Universe contained enough mass, it would be curved into a sphere of four dimensions, rather than three. The three dimensions that we occupy would be confined to its surface. Light traveling in any direction around this sphere would follow a circular path, like a satellite in orbit.

Blaise Pascal was one of the great mathematicians of all time. As a teenager he published original mathematical theorems and designed a mechanical calculator. His name is immortalized in the computer language PASCAL. In the 17th century Pascal wrote, "Nature is an infinite sphere of which the center is everywhere and the circumference nowhere." Pascal also thought that the Universe was spherical.

In the 'hood of Baltimore, in front of a housing project, sits an unassuming little house. The mind that lived within these brick walls gave us *The Raven, The Fall of the House of Usher*, and *The Pit and the Pendulum*. 203 Amity Street was the house of Edgar Allan Poe. While physicists see the pendulum as a simple harmonic oscillator, Poe saw the terrifying, inexorable approach of death.

After *The Raven* brought fame and success, the heart of Poe's beloved wife stopped beating in 1847. Poe himself would pass on just 2 years later, living only slightly longer than Pascal. Possibly sensing the pendulum's approach, Poe in 1848 finished his favourite work. Of all his writings, poetry and prose, Edgar Allan Poe was most proud of *Eureka*. It was a poem not about death or suffering, but the origin and evolution of our Universe.

Like Pythagoras and Pascal, Poe thought a sphere the most natural shape. Most prescient, Poe suggested that our giant Universe had expanded from a single tiny point! "From one particle, as a center," Poe wrote, "let us suppose to be irradiated spherically—in all directions—to immeasurable but still to definite distances in the previously vacant space." He thought that our galaxy was just one of many. The master of dark fiction also predicted that the Milky Way's centre contained a massive unseen object, which he called a "non-luminous sun." Predicting a Black Hole at the centre of our galaxy, Poe's *Eureka* was far ahead of his time.

Albert Einstein concluded that we live in such a sphere. The local conditions of Special Relativity, which assume a flat Space with no gravity, apply in our small patch of the Universe. If a starship could travel much faster than light, speeding off in any direction would eventually lead back to where it started. Speeding off in an entirely different direction would return it to the same point. (In an episode of *Star Trek:: The Next Generation* the crew of the starship *Enterprise* find themselves trapped in a spherical space. They leave a buoy behind to mark their starting point. Any direction the *Enterprise* warps to returns them to the buoy.)

Though curved on the large scale, for small distances Einstein's Universe may be considered flat. From tiny raindrops to planets, large objects under the influence of gravity tend to form spheres. The Universe is very large. Since long before Columbus, the "Flat vs. Curved" debate has raged on Earth. Everyone knows which side wins.

The Universe imagined by Einstein has no centre, but every bit does feel an attraction from the whole. Newton's calculus had shown 250 years before that the gravitational attraction from a spherical shell is the same as is all the mass were in the centre. The combined gravity of the Universe would pull each point inward. Newton's mathematical proof applies in 3 or 4 dimensions.

Here Einstein encountered a conflict. The same gravity that curved Space/Time into a sphere would cause that Universe to collapse. We add here: Gravity would cause a spherical space to collapse…unless it was expanding. The momentum of expansion would prevent the Universe from collapsing. If Einstein had proposed an expanding Universe it would have been one of the great scientific predictions in history. At first others would have ignored this prediction, for astronomers took 15 years to discover expansion.

"A long time ago in a galaxy far, far away" is today part of our lore, but in 1917 our Milky Way was the

87

known Universe. Astronomers did not agree how far its disk of stars extended, where its centre was located, or even if the Milky Way had a centre at all. They disagreed whether "spiral nebulae" like Andromeda were part of the Milky Way or other island galaxies. Though the spectra of these objects showed a shift into red, other data seemed to show that the Universe was unchanging. Einstein hoped to satisfy a Perfect Cosmological Principle, where the Universe looked the same everywhere in both Time and Space. Possibly motivated by a search for harmony, Einstein believed his sphere to be static and unchanging. In a static Universe, the speed of light would be constant.

A spherical balloon would collapse without something to support it. Something had to prevent the static universe from collapsing. Here Einstein made his biggest mistake. He added to his equations a repulsive term, a "cosmological constant." This anti-gravity force would fill Space/Time and oppose the tendency to collapse. Like epicycles and ether, the cosmic constant would be invisible yet fill all of Space. Einstein provided no theory to explain the constant, and no experiment could isolate it. The constant was a fudge factor added to keep the spherical model from collapsing. The cosmological constant completed Einstein's Static Universe.

Einstein wrote about his static Universe in 1917, on the brink of fame. A solar eclipse in May 1919 would provide another opportunity to test General Relativity. British astronomer Arthur Eddington ventured to South America to observe the eclipse. He returned proclaiming to the press that Einstein's Relativity had been proved. Though the evidence today seems inconclusive, Eddington made the world take notice.

On November 7, a year after Kaiser Wilhelm's abdication, the London *Times* announced, "REVOLUTION IN SCIENCE." To a world weary from the noise of war, the quiet Einstein became an instant celebrity. In the following year over 100 books appeared about his theory, including Einstein's own *Relativity*.

The rise of his fame began a decline in Einstein's scientific output. His great discoveries involving Relativity and the photoelectric effect were behind him. The demands of fame took up much of Einstein's time. Most of his contributions to science would be judging the work of others. In a famous letter to President Roosevelt, Einstein warned of the possibility of an atomic bomb. The job of expanding upon his discoveries fell on other shoulders.

1924 was a very good year for waves. Hawaii's great surfer Duke Kahanamoku won a swimming medal at the Paris Olympics. On the other side of Paris at the Sorbonne a Duke named Louis De Broglie made waves with his doctoral thesis. Maxwell had shown that light was an electromagnetic wave. Duke De Broglie suggested that the matter we are made of takes the form of waves, dependent on Planck's value h.

Originality can be bad for an academic career. De Broglie's thesis was so short, simple and important that his professors were ready to reject it. Fortunately De Broglie was an aristocrat, and his older brother prevailed upon the school to get an outside opinion. Albert Einstein recommended that the Sorbonne give De Broglie his PhD, and no one dared contradict Einstein. 5 years later Einstein would nominate De Broglie for the Nobel Prize—Einstein was a very good friend to have.

Among the eyes who saw the thesis was physicist Erwin Schrödinger. With De Broglie's paper in hand, Schrödinger spent 2 weeks in a Swiss ski resort with a woman friend. After consulting her repeatedly, Schrödinger produced equations for the wave function of matter. The work of De Broglie and Schrödinger showed in detail that particles of matter could also behave as waves, with an energy given by Planck's h. Connecting this quantum world with the Universe of Relativity is one of the great problems in physics.

Isaac Newton showed that light waves could be broken up into spectra. Different elements gave off different spectra, and quantum mechanics provided a way of calculating those spectra. The "spiral nebulae" showed spectra that were shifted into the red, as if these objects were receding. Many theories were proposed to account for this redshift, but the answer lay in an expanding Universe.

Alexander Friedmann was born in St. Petersburg, Russia to a musical family. His mother was a pianist, his father a ballet dancer and composer. While studying physics, he became interested in applications to meteorology. During the First World War, he volunteered to use his knowledge to aid the new tactic of dropping bombs from planes. He flew as an observer on many hazardous missions, becoming known to both sides. On days when Russian bombs were on-target, the Germans would mutter "Friedmann is in the air today."

After the war, despite the distraction of the Russian Revolution, Friedmann devoted himself to the consequences of Relativity. He also believed in the cosmological principle that the Universe looks the same in every direction and that every bit resembles every other bit. Friedmann found solutions to Einstein's equations that indicated an expanding Universe, doing away with the need for a cosmological constant. Unfortunately Alexander Friedmann passed away on 1925.

Meanwhile a Belgian priest named Georges Lemaitre had both God and the Universe in mind. After attaining the priesthood, he began graduate study at Cambridge University, with Arthur Eddington among his teachers. In a 1927 paper Lemaitre proposed an expanding Universe, with the additional proposal that expansion caused redshift of nebulae. At a conference that year Lemaitre approached Einstein with his idea. The famous scientist could find no flaw with the mathematics, but rejected the idea of expansion. Einstein also pointed out that Friedmann had come up with a similar idea. It would be several years before Einstein accepted that Lemaitre was right.

During 1923-24 astronomer Edwin Hubble, working atop Mount Wilson in California, made many observations of "spiral nebulae." His distance measurements proved that they were other galaxies far outside the Milky Way. There was nothing special about our location in space--our galaxy was just one among countless others. Thanks to Hubble, the galaxies became far, far away.

At Mount Wilson, Hubble and his colleague Milton Humason were carefully observing and cataloguing the galaxies. They relied upon the light of stars called Cepheid Variables. These stars vary periodically in brightness, with a period related to their luminosity. Observing a Cepheid's period would tell them how much light the star gave off. By measuring how much of that light reached Earth, they cold determine distance to the star and galaxy. Cepheid Variables were standard candles measuring the distance to their galaxies.

Astronomer Henrietta Leavitt had previously discovered the relationship between Cepheid periods and their luminosity. Leavitt became interested in astronomy during her senior year in university. She worked as a volunteer assistant at the Harvard Observatory for seven years before being hired for 30 cents per hour. Her supervisors did not want a woman doing theoretical work, so Leavitt examined photographic plates and maintained telescopes. Despite this early challenge, she made many discoveries about the magnitude of stars, including the use of Cepheid Variables as distance markers. Hubble's discovery of an expanding Universe would not have been possible without Henrietta Leavitt.

When an object is moving rapidly away from us, its light is redshifted. This is similar to the Doppler effect, which causes the pitch of a sound to change when the object is moving toward or away from the listener. The light from a receding object is shifted into longer wavelengths, white light would be shifted into red. Redshift is approximately v/c, an object's velocity divided by the speed of light. When the redshifts of many galaxies were plotted against their distances, they increased linearly with distance. This did not indicate that our galaxy was unpopular. If the Universe was spherical like a balloon, the galaxies were like spots on its surface. As a balloon expanded, the spots would increase their distance uniformly. A galaxy twice as distant would recede twice as fast. The distance-redshift relation was convincing evidence that our Universe was expanding.

In a celebrated 1931 visit to Mount Wilson, Einstein conferred with Hubble and peered through the telescope. The light in Hubble's telescope convinced Einstein that the Universe was expanding, forcing him to drop the cosmological constant. Later Einstein would call the constant his "greatest blunder," for it prevented him from predicting an expanding Universe.

I recently consulted with the Einstein statue in Washington. On the tablet in Einstein's hand are his famous equations, minus the cosmological constant. Einstein's statue, never one to blindly follow fashion, has not seen fit to re-introduce the constant. No one today should repeat this constant, for if Einstein called something a blunder he was probably right.

In a final contribution to cosmology, Einstein co-authored to a 1932 paper with Willem de Sitter. In this paper they both favoured a cosmological model that expanded and slowed asymptotically. This Einstein-de Sitter Universe expanded at the rate of its age t to the 2/3 power. Mathematically it is the simplest of models. Its expansion continually slows, but will never slow to a stop or reverse. Density of this Universe is a "critical" value called Omega or Ω, neither too light nor too dense. Einstein and De Sitter also suggested that large amounts of mass remain undetected. Einstein-de Sitter was long the cosmologist's favourite model.

Einstein's early work was prodigious, but all things slow with time. Einstein spent his twilight years as the world's most famous scientist, but personally frustrated. He spent years trying to find a Unified Field Theory to explain the Universe simply. He did leave us one final means of judging ideas. When asked what he would think if a theory of the Universe was complicated, Einstein replied, "then I would not be interested in it."

In 1948 George Gamow made one of the most important predictions of cosmology. If the Universe had

95

once been smaller, it would have been dominated by radiation. The Big Bang would have been very hot. Gamow predicted that the radiation should still be around, coming from every point in the sky. Gamow greatly overestimated the temperature at 50 degrees Kelvin. If the temperature were known, Planck's blackbody formula could predict the spectrum of that radiation. Because Gamow's work was largely ignored, the two physicists who discovered this radiation initially had no idea what they had found.

In 1965 Arnold Penzias and Joseph Wilson, working at Bell Laboratories, were working on a microwave antenna for communicating with satellites. As with all radios, there was a background of hisses and pops to be removed. After checking everything including bird droppings, the two physicists could not account for a mysterious background signal. This radiation appeared to come from every direction in the sky. The temperature of this radiation was 3 degrees Kelvin, close to predictions of expanding Universe models. Discovery of the CMB was "smoking gun" evidence that the Universe evolved from a hotter, denser state.

Aristotle would marvel at the amount of data scientists collected. As evidence mounted in favour of the expanding Universe, there was still no guiding principle.

96

The master Plato would ask, why does the Universe expand? An answer could tell us both how it began and whether it will end.

A Simple Principle

In imagining the spherical Universe, Einstein was far ahead of his contemporaries, as far as was possible given the observations of his time. Today we have far more evidence to work with. The Universe appears to have a beginning, commonly called a "Big Bang." Near this time of great heat and density, an enormous mass was confined in a very small Universe.

This beginning can be drawn as a point. Cosmologists agree that this baby Universe had a finite size. Therefore it had a finite mass M.

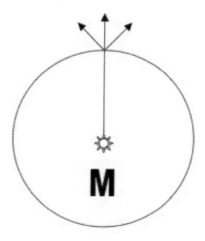

Near the Big Bang, all points of this Universe were very close to one another. Our separation from the Big Bang is a matter of Time, not Space. Astronomers have estimated that age t is about 13.7 billion years.

Our light cone represents the local conditions of Special Relativity. Our mass, even the mass of our galaxy, is negligible compared to M of the Universe. We are within the light cone of that enormous mass, and its gravity affects us. As Newton showed, the gravitational force from a spherical mass distribution is the same as if all that mass were concentrated in a point.

This drawing, simple as a child's thought, represents the entire Universe. Our three dimensions x, y and z are again compressed into the page. This Universe has a spherical shape. Any direction that we can travel in Space keeps us in that narrow circle. The Universe might appear infinite and flat to our experience, but still be curved in the fourth dimension. From a sphere, the combined gravitational attraction is the same as if everything was in a central point. There is no centre in Space, for every bit resembles every other bit. There is a centre in Time, which we call a "Big Bang." As time t increases, our Universe expands away from the Big Bang.

Newton's Law of Gravitation insists that mass affects us at a distance. It is meaningless to express this attraction

over Time, so we must add the conversion factor c. Here we expand on one Principle of Special Relativity.
Since Space and Time are one phenomenon related by c: Scale R of the Universe is its age t multiplied by c.

Near the Big Bang the Universe was tiny--all points within it were from our point of view the same distance away. Our separation from the Big Bang's centre is simply the age t of our Universe. Expansion of the Universe is then indistinguishable from the forward flow of Time. This simple principle tells us why we live in an expanding Universe: As time t increases, scale R expands.

A simple principle also explains an "Arrow of Time." Most laws of physics are time-symnetrical; they apply equally forward or backward in Time. Expansion does not share this symmetry. As Time increases, the Universe only expands outward. Scientists call this the Cosmological Arrow.

Since that distant time of the Big Bang, our Universe has been expanding. It can't expand at the same rate continuously, for gravity slows it down. Science would not be complete if it did not make predictions. The speed of light c must be further related to time t. This leads to one equation with many predictions.

NEXT: The Only Equation In This Book

6

<u>The Only Equation in the Book</u>

A Simple Thought

Mathematics can be both beautiful and puzzling. A truly beautiful equation can describe many phenomena with a single line. Equations can also frighten people completely away from science. Many budding scientists and engineers have abandoned those dreams after glimpsing the amount of math involved. The mathematics of epicycles was incomprehensible except to a learned elite. Many layers of complexity kept the theory of invisible spheres unchallenged for centuries. The most lasting equations are simple.

This book contains just one equation. A good authority claims that every equation in a book cuts the number of readers in half. It would be tempting to write a book with no equations at all. Such a book could make grand promises about theories of everything, without offering anything that could be proved by experiment. While grand promises can be good for careers, the purpose of science is to describe the world in simple rules.

Since there is only one equation in the book, it should describe as much as possible. It should be simple enough for the young people reading this book to grasp. Including an equation from anyone but this author would be quite improper. I will omit the many years of calculations to reach this equation. It is pictured with a spherical Universe similar to that of Pascal, Poe and Einstein.

"When forced to summarise the General Theory of Relativity in one sentence," Einstein said, "Time and Space and Gravity have no separate existence from Matter." Today we can translate that statement into an equation:

$$GM = tc^3$$

Where G is Newton's gravitational constant, M is mass of the Universe, t is its age and c is the speed of light.

The drawing and equation fit on a T-shirt, yet tell us everything we wanted to know about the Universe but were afraid to ask: How big it is, how fast it expands, and whether it will ever slow to a stop. This allows us to make predictions about the density of mass, the proportions of different kinds of mass, the redshifts of distant supernovae,

102

and many other phenomena. It is pleasing when our big and mysterious Universe can be described simply.

We can describe the equation bit by bit. Newton's gravitational constant G has so far appeared to be constant. Arthur Eddington and Paul Dirac both suggested hypotheses of varying G. These theories were disproved by many experiments, including the first spacecraft to land on another planet. Data from the Viking Lander showed that Mars' orbital period was not changing, as it would if G varied. The best evidence for a constant G is the existence of life on Earth. If Newton's G changed with time, Earth's orbit and temperature would not be stable enough for life to evolve.

Since this Universe has a finite size, the mass M contained within is immense but also finite. This quantity M may change slightly, for small amounts of mass are created or destroyed in nuclear reactions. During the hot times near the Big Bang, this may have happened quite often. In its present cool state, the Universe has a nearly constant mass. The Universe is big—but we can envision how big from pure math. Since G and c can be measured in a lab, and age t can be estimated from astronomical observations, we can figure out how massive everything in the Universe is. That mass M is over 100 billion times the mass of our Milky Way galaxy. Since G and M are nearly constant, both sides of the "equals" sign are also constant.

The dimension of Time is represented by t. At one time people thought that this age was only a few thousand years. Only in fairly recent history have we known Earth and the Universe to be billions of years old. Einstein at first thought the Universe was eternal and unchanging, leading to his blunder of a cosmological constant. Today evidence from redshifts to the microwave background all indicate that the Universe has a finite age. At this writing, the best estimate of age t is about 13.7 billion years.

Three dimensions of Space are represented by c, the speed of light. As we have seen, c is a conversion factor between Space/Time. Since both sides of the 'equals' sign are constant, as t increases c slows down. When t was tiny, c was enormous, and the Universe expanded like a "Big Bang." As t increased, that expansion slowed due to gravitation, and continues slowing to this day. Now that t is billions of years, change in c is very small. Though today's speed of light is 300,000 kilometres per second, in one year it would change by only 0.72 centimeters per second! A simple equation allows us to predict the value of c in the past, present, or future.

First we can solve for the scale R as a function of time. The expansion rate turns out to be proportional to $t^{2/3}$, as Einstein and de Sitter believed. The Universe expands and slows asymptotically; it will never stop, reverse or collapse. This tells us that there is nothing particularly special about the time we live in. The Universe has been behaving according to $GM = tc^3$ since it was no bigger than the drawing, and will do so indefinitely.

As Einstein believed, the Universe follows the Perfect Cosmological Principle. Every bit if the Universe resembles every other bit. Not only does it follow the Principle in three dimensions, but in the fourth dimension of Time. The Universe looks more or less the same at any time. Surprisingly, the "Big Bang" has not ended. We are still in it, part of a continuing process of expansion and slowing.

When scale R is plugged into Alexander Friedmann's equations, the "critical" density Ω of Einstein and De Sitter pops out. This is not just a critical density but the most stable density. If the Universe were below this density, quantum theory predicts that matter would form via pair production. In this process, particles and anti-particles would form until the stable density Ω was reached. The Universe has exactly the density to continually expand and slow.

The stable density was not the initial density. Before any matter formed, the Universe was underweight. The matter we are made of—protons, electrons, and neutrons, would have formed to fill this gap. The difference, between density before matter and density with matter, can be calculated from pure math. That proportion is approximately 4.507034%. The matter that we are made of is just that small slice of the Universe.

The First Law of Thermodynamics states that energy is conserved—it cannot be created or removed, just changed into different forms. Every object before our eyes contains energy. Newton showed that objects have both potential and kinetic energy. Einstein added a rest energy given by his famous equation. When all these energies are added up, the total is Zero! This result applies to any object, from the tiniest particle to the heaviest galaxy. Even photons, the massless particles of light and electromagnetism, have the same total energy. Our enormous, complex Universe has zero energy. It is the ultimate free lunch, which has allowed it to expand from a tiny point to an enormous size.

Humans may have first suspected that Earth was spherical because of objects disappearing over Earth's horizon. When the speed of light is known, the distance our eyes can see can also be calculated. Scientists call this the cosmic horizon, the distance light has traveled since the Universe began. This turns out to be exactly (3/2) times the radius. On Einstein's sphere that is almost halfway around. If our position were the North Pole, we could see nearly to the equator. Light that reaches us from near the equator dates from a time near the Big Bang. The cosmic horizon expands at the same rate as the rest of Space/Time.

The Moon keeps the same face turned towards Earth. People often refer to "The dark side of the Moon," a wonderful title for a Pink Floyd song. The far side of the Moon isn't dark--it receives just as much sunlight as the near side—but it was hidden from human eyes until we could send spacecraft to observe it. Like the Moon, the Universe has both a visible side and a hidden side that is forever beyond our sight. The Moon offers some of the best evidence that light is slowing down—more about that in Chapter 8.

If the speed of light c did not change, the visible patch would grow bigger. Astronomers' lives would be even more exciting, for they would see distant objects suddenly appear in the sky! After a long enough time, our telescopes would see completely around the sphere to view ourselves. Such a Universe could not support life, for the fiery radiation from its beginning would make its way around the sphere to cook us. Fortunately for our existence, light slows at just the right rate.

c *and h*

If the speed of light c varies, something else must change. Planck's value *h* is critical to quantum mechanics. It allowed Max Planck to explain the blackbody spectrum, and Einstein to explain the photoelectric effect. It is also a key to the Uncertainty Principle, which says that particle positions cannot be measured exactly. Speed of light c and *h* are multiplied together in the fine-structure constant α. This constant is important in predicting spectral lines from stars like the Sun. Quantities c and *h* also appear together in Chandrasekhar's Limit, which relates to exploding supernovae. Because our Sun's mass is below this limit, it will never explode into a supernova. As Einstein found, c and *h* appear together in the energy of light particles.

Most laboratory measurements indicate that the product *hc* is indeed constant. This is not an obstacle, but

108

a possible link between Relativity and Quantum Mechanics. If c is slowing but the product hc is constant, then Planck's value h increases. The quantity h is in fact the solution to a random-walk problem in three dimensions, such as physicists use for subatomic particles. Directly measuring change in h would be even more difficult than for the speed of light, for Planck's value is only revealed in the microscopic world. Ole Roemer roughly measured the speed of light back in 1676, but not until 1900 did Planck measure h. Many undergraduates struggle to remember the Planck value. Fortunately most students know the present value of c. It is an easy step to remember hc, which is just 2×10^{-25}. If a student divides by c, she will get Planck's value h correct to two decimal places. In this small way, the big Universe of Relativity and the microscopic quantum world may be linked.

Planck's value h is a key to the other arrow of time, the thermodynamic arrow. Nature is full of processes that are irreversible. If a plate is dropped from a high place, it shatters into many pieces but never reassembles itself. Heat from a fire spreads throughout a room but does not converge again at the fire. These are demonstrations of entropy, the tendency for disorder to increase. These thermodynamic processes are governed by the value h. An increasing h means that the entropy of the Universe

109

increases with time. Its complexity, and the creation of structures that lead to life, also increases.

In the Beginning: Initial Conditions

From childhood we have asked where the Universe came from. This question is the basis of cosmology and many other sciences. Every culture has created legends to try and explain how the Universe was created. Today we can imagine some initial conditions for how the Universe began. These initial conditions led to the complexity we observe today.

In the beginning the Universe was indeed without shape or form. There was no Space or even Time. From the perspective of this beginning, every direction was forward in Time, like looking outward from a single point. Anything before this is beyond human imagination, for not even Time itself existed. The Universe began at a time of Zero.

At time Zero the Universe also had zero radius, like a point. The Planck value h was also zero, for with zero scale there is no uncertainty in position of anything. Creation began with the speed of light. When time t was near zero, the speed of light c was nearly infinite. This immense speed caused the Universe to expand like a Bang. As today, the total energy of the Universe was

zero. There was no need for any additional energies--
these initial conditions were all that was needed to cause
the Big Bang.

Collapse and Creation

The Universe could not explode at such an immense
rate for long--gravity almost immediately started to slow
expansion. The Planck value h also grew, though not as
fast as the Universe. Though the overall energy was zero,
quantum uncertainties ensured that the Universe did not
have the same density everywhere. These fluctuations
were magnified by the rapid expansion. Some of these
density fluctuations grew so large that they collapsed due
to their own gravity. Gravity from these Black Holes
sucked in the mass around them, clearing great voids in
Space. Almost nothing could exist in these voids except
for the huge masses in their centres. As the Universe
continued to expand, some of these voids grew to
immense size.

Primordial Black Holes formed in a variety of sizes.
Large Black Holes would someday seed formation of
voids, clusters, galaxies and even smaller structures.
Billions upon billions of tiny Black Holes were also
created. Formed shortly after the Big Bang, they have
never been matter, and would be nearly invisible to our

eyes. Drawn by gravity toward larger objects, they would form dark haloes around galaxies.

The matter that we are made of formed next. After the first Black Holes had formed the Universe was still underweight by 4.507034%. The Universe was very dense and hot with radiation from billions of Black Holes. Quantum mechanics predicts that this Space was filled with virtual particles winking in and out of existence. Enough of these matter particles remained to bring the Universe up to its stable density. This matter would form stars, planets and eventually life. The first atoms formed within minutes of the Universe's beginning. Those details are beyond the scope of a small book. The wonderful complexity of matter we create every day.

The language of mathematics can be quite powerful. A single equation may describe an entire Universe. The most beautiful equations would be useless to our understanding if they did not make testable predictions. Science is just speculation if it is not supported by results. Extraordinary claims require extraordinary evidence. Next we will see the results of many experiments.

NEXT: The Proof Is In the Pudding

7

The Proof In the Pudding

Supernova image courtesy NASA

The Universe has been compared to raisin pudding, where the raisins are galaxies. As the pudding expands, the individual raisins grow further apart. A raisin twice as far away recedes twice as fast. The more distant an object, the faster it recedes from us, causing its light to be redshifted. We are somewhere within that pudding, looking for clues. The duty of scientists is to make predictions that can be verified by experiment.

113

Like the Renaissance or the beginning of the 20th century, today is an exciting day to be in science. Spacecraft have extended humanity's reach into the solar system. New discoveries and data arrive almost daily. Like a baby's first months, observations come so fast that scientists don't know what they are seeing. Hopefully this book can convey the thrill of discovery.

In case the reader has skipped the last chapter, the Only Equation in This Book predicts that light is slowing down at a very tiny rate. Science should make testable predictions so that theories can be adopted or discarded. Change in c is not a new idea. In 1874 William Thomson, the First Lord Kelvin published a paper claiming that light slows down. In recent years, a growing number of physicists have questioned whether the speed of light has always been the same.

Here are described data from many spacecraft and experiments. Several lines of evidence support the prediction of a changing speed of light. In each case, alternate explanations for the data are mentioned. For science to advance, ideas deserve a fair hearing. A first colourful picture comes from shortly after the Big Bang.

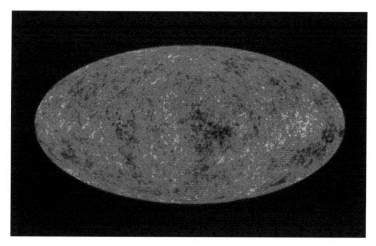

Cosmic Microwave Background courtesy WMAP

The Cosmic Microwave Background

This is light from near the Big Bang, predicted by George Gamow. Our picture of the Cosmic Microwave Background comes from the Wilkinson Microwave Anisotropy Probe (WMAP). This spacecraft orbits at the L2 point, on the other side of the Moon. Its mission is to make detailed temperature maps of the background sky. The COBE spacecraft and BOOMERANG balloon experiment made previous maps of this background. Penzias and Wilson's accidental discovery was radiation from 380,000 years after the Big Bang. Expansion and Time have stretched its wavelength and cooled its temperature to just 2.7 degrees above absolute zero.

The most striking feature of the microwave background is its uniformity. The temperature is uniform across the sky to one part in ten thousand. Colours in the picture represent very tiny fluctuations. If we look at this picture, the average temperature is the same over large parts of the sky. This indicates that large regions of the Universe were within the range of light signals. Even at a time barely 380,000 years after the Big Bang, the speed of light was much larger than today.

In the late 1970's, when rising prices were on everyone's mind, the paradigm of inflation was proposed. Physicists assumed that the speed of light was always the same, but when the Universe was just 10^{-33} seconds old it suddenly expanded at warp speed, many times faster than light. The inflation idea has been a useful step, but it is incomplete as a theory. Inflation would violate both the First Law of Thermodynamics, which states that energy is conserved, and Relativity's stipulation that nothing can travel faster than light.

Inflation relies on hypothetical repulsive energies to break these laws. These "inflatons" or "scalars" would have appeared at the Big Bang, yet conveniently disappeared before we could observe them. While inflation of prices has since been tamed, the inflationary paradigm has persisted. It has spawned an inflation of different theories, but a mechanism is still not known. Scientists cannot time-travel to the first 10^{-33} seconds to observe inflation. No conceivable experiment can approach the titanic energies near the Big Bang. Since nothing like an "inflaton" has ever been observed in nature, inflation may never be proven.

The inflationary paradigm states the infant Universe abruptly expanded to 10^{43} times its original size! While it began tiny and spherical, it expanded so large that it now appears flat. Quantum fluctuations in density would expand to almost infinite size. Using the equations of spherical harmonics, theorists can predict the spectrum of these fluctuations.

Results from two satellite experiments, WMAP and COBE both show a spectrum far different from inflation's prediction. Their data falls outside the error bars. Most striking, the graph is nearly zero for angles greater than 60 degrees. Inflation with a flat universe predicts that they should extend to infinitely high angles. As a ship disappearing over the horizon shows that Earth is spherical, the cutoff indicates that the Universe is also curved. Large objects, even raindrops, tend to form spheres. The Universe is very large.

The principle from Chapter 5 states that Scale of the Universe is its age multiplied by c. Any changes in density would also be of limited size. When Theory's prediction is plotted on the graph, it fits the data precisely. The close fit indicates that the Universe is curved with the predicted radius. When prediction fits the data this closely, it may mean something! Though inflation has been a useful placeholder, WMAP scientists concluded that an alternative model would be an exciting development.

Light from the cosmic background provides a huge amount of information. Peaks in the graph indicate the overall density of the Universe, and even proportions of different kinds of mass. A first peak indicates the overall density. That turns out to be exactly the "critical" density for Einstein-de Sitter expansion, which is also the stable density predicted by Theory. Succeeding peaks indicate the amount of density that is the matter we are made of. This is predicted by Theory to be 4.507034%; WMAP's result is $4.4 \pm 0.3\%$. The amount of mass that would be invisible, having been sucked into giant Black Holes, can be predicted at about 68%. This is also in line with data from the CMB.

The huge amount of microwave data allows many interpretations. It can be fashionable to say that the Universe is flat, like the Earth. Flatness can appear to be supported by data. A typical distance between peaks is about one degree. The microwave background dates from a time very near the Big Bang. The distance light has traveled since that time was assumed to simply be ct,

118

age of the Universe multiplied by the present speed of light. Combined with the distance between peaks, this gives three sides of a triangle. The angles total 180 degrees, seeming to prove that this triangle was on a flat space.

The triangle example, seemingly solid as Pythagoras, does not consider that the speed of light slows down. The distance light has traveled is not ct, but $(3/2)ct$. The angles do not add up to 180 and the triangle is not flat. The greater distance indicates that the Space is curved, and how much.

Again we take two peaks separated by one degree. Instead of on a flat map they are near the equator of a globe with radius ct. Our location would then be on the North Pole, nearly 90 degrees away. If we draw curved lines out to those two peaks on the equator, they will appear one degree apart. They will only appear one degree apart of the Universe is of radius ct. Including the change in c, the data shows that the Universe is curved with this radius. Even inflation advocates admit that the Universe must have curvature. It is not mathematically possible for the Universe to expand from a tiny point into an infinite flat plane. Such a universe would also have infinite mass, and could not expand at all.

Unlike inflation, a proper theory makes a testable prediction that the speed of light is slowing. The microwave sky is strong evidence for change in c. Data from this background can be interpreted in more than one way. Inflation with flat space can be fitted to observations, but is ruled out by the scale of density fluctuations. A spherical Space/Time makes an even more precise fit to data, and also predicts the densities of mass. Signs of c change come from huge objects not quite as distant.

Galaxies

Every galaxy, like our Milky Way, contains at its centre a massive Black Hole. Giant Black Holes very likely exist in Space without surrounding galaxies. Instruments like the Hubble Space Telescope and the Subaru Telescope atop Mauna Kea can find galaxies that existed only a few hundred million years after the Big Bang. They are so distant in Space/Time that their light has shifted into the infrared. In the distant past we have seen quasars and Active Galactic Nuclei, powerful objects each radiating more light than a million galaxies. These primordial objects are powered by massive Black Holes.

Near the beginning of the Universe, not enough time had passed for Black Holes to evolve from stars.

Primordial Black Holes could have formed by collapse of quantum fluctuations. In the immense densities near the Big Bang, fluctuations were big enough to form Black Holes without becoming matter first. Many scientists, notably Stephen Hawking, believe that these Primordial Black Holes survive today.

Again scientists did not consider whether the speed of light had changed. The mass of a collapsing Black Hole is limited by a horizon distance, the distance light can travel in a given time. Previously it was thought that Primordial Black Holes would be relatively tiny, because of light's limited reach. If we consider that the primordial speed of light was faster, the mass within the horizon was enormous. Primordial Black Holes could have formed of immense size, seeding formation of galaxies around them. The massive Black Hole at the centre of our Milky Way may be older than the galaxy.

Though we live in the Milky Way and see its band of stars at night, astronomers do not understand how our galaxy formed. It had been thought that the Milky Way formed from the merger of smaller galaxies. An international team led by astronomer Manuela Zoccalli has disproved that inference. Her observations show that stars in the galaxy's central bulge have a different chemical composition than stars in the spiral arms. If the

Milky Way had formed from collisions of smaller galaxies, stars of similar ingredients would be mixed together everywhere we look. Stars from the outer arms would have been pulled toward the centre. Zoccalli's observations also show that our galaxy's central bulge formed in an amazingly short time, less than a billion years.

Observations from the Subaru Telescope atop Mauna Kea have found galaxies formed barely 500 million years after the Big Bang. As we look farther into the past, we continue to see signs of massive Primordial Black Holes. These objects could not form without a higher value of c. If these objects date from near the Big Bang, they are indirect evidence that c was once much higher. If c had not been higher, our Milky Way galaxy could not have formed. More precise and controversial evidence comes from the brightest explosions in galaxies.

Supernovae

When a star explodes, its light can briefly outshine an entire galaxy. When Tycho Brahe saw a supernova in 1572 he could only speculate what it was. A debt is acknowledged to observers who find exploding stars while stumbling to interpret what they see. The true meaning of their work may someday be recognised.

122

Though the power source of these huge explosions has not been understood, enough is known from observations to make predictions about their behaviour. Supernovae have been classified into various types. Physicists have focused on Type Ia supernovae, which have been observed in many galaxies and appear to all have the same luminosity. From observations, predictions have been made as to their spectra and brightness.

Edwin Hubble used Cepheid Variable stars to show that galaxies were receding. Though their power source was not understood, Type Ia supernovae were considered as standard candles. By searching for supernovae at high redshift, observers hoped to extend the Hubble diagram further into the past. Redshifts of nearby supernovae increased in a straight with distance, indicating that the Universe expands as Friedmann, Einstein and de Sitter predicted.

When observations were extended to high redshifts, the line curved mysteriously upward. Redshifts, which are dependent on velocity v divided by c, appeared to have accelerated. The lead authors of the two "independent" groups were both working from the same campus, so they could check their independent findings. The lead author of one group's paper had an office in Campbell Hall at University of California Berkeley, the other group's lead author was on the same campus at Lawrence Berkeley Laboratory. It benefited them to be in two teams, for a discovery made by a single researcher might not have been accepted.

Two large observing teams did not consider whether speed of light c had affected their measurements. They concluded that velocity v and the Universe was accelerating. Because acceleration would violate the First Law of Thermodynamics (Conservation of Energy) Einstein's "blunder" of a cosmological constant was revived. Since data showed that the cosmological constant could not be constant, it was renamed as "dark" energy. Like epicycles and ether, this energy would be invisible yet fill most of Space. Some researchers dug up Aristotle's millennia-old idea of a "quintessence". Unlike all other forms of mass and energy, which attract each other gravitationally, "dark" energies would exert a repulsive force. No experiment could isolate this invisible actor; there is not a single sample or track in a bubble chamber to prove its existence. "Dark" energy is often called a fudge factor.

Redshifts of supernovae are the only evidence of "cosmic acceleration." The microwave background, dating from a time only 380,000 years after the Big Bang, tells us nothing about acceleration. The supernova observers did not know what they were seeing. (At this writing, they've still not figured it out.) Though their observations were highly valuable, two large groups of scientists did not consider what a child could ask: What if v/c increases not because v accelerates but because c slows down?

Not far from the Einstein statue in Washington DC is the monument to another Nobel Prize winner, Dr. Martin Luther King.

On the wall behind the memorial is a quote from Dr. King:

> "Darkness can not drive out darkness,
> only light can do that.
> Hate can not drive out hate, only love can do that."

The answer to accelerating redshifts may not be in "dark" energies but in the slowing of light.

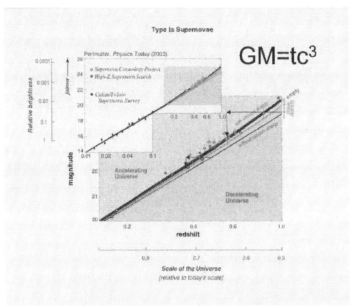

Courtesy of Supernova Cosmology Project

An object of redshift 1 recedes at 60% of today's speed of light. That would be only 42% of c at the time that light was emitted. The redshift we observe will be only 0.57. (Horizontal arrow) Supernovae convert mass m into energy E by that famous Einstein equation with mc^2. For a supernova of redshift 1, that energy would be doubled, changing its magnitude by -.75. (Vertical arrow) The prediction line is hard to see, because it is covered with data points. Given the observer's error bars, the prediction intercepts 95% of the points. The "acceleration" curve can be precisely predicted from light, without imaginary energies. Every point on the graph indicates that light has been slowing down.

126

It is still possible to profess that the Universe accelerates because of some repulsive "dark" energy. If scientists turn to that dark force, they may follow its dark path for years or decades. A complicated equation of state for "dark" energy can also be produced to match the data. Just as enough epicycles can put Earth at the centre of the cosmos, they can make the Universe appear to accelerate.

Speculations of "inflation" and "dark" energy lead not to a solution but to a divergence of theories. Both mysteries may be explained by a slowing speed of light. To verify that the speed of light has changed over time, corroborating evidence is required from truly independent sources. That may come from the nearest star of all, and from our Moon. The proof is in the pudding.

Next: Sunlight and Moonlight

127

8

Sunlight and Moonlight

A Hot Young Sun

Before the invention of fire, the Sun and Moon were our only sources of light. The Sun greets us each morning. "Sunrise" is of course a misnomer; the Sun only appears to rise because of Earth's rotation. Though humans today know that Earth orbits the Sun rather than vice versa, the term has stuck. Our eyes see the world thanks to light from the Sun. Heat from the Sun's radiation has allowed life to evolve on Earth's surface. The Sun and its light have been sources of awe, mystery and even worship. The Sun and its life-sustaining heat tell us something about the speed of light.

Scientist George Washington Carver enjoyed the light of sunrise. He rose every day before dawn to walk through the woods. The plants and crops grown in the Sun's light fascinated him. Among other innovations, he invented nearly 100 products made from peanuts. Life on a planet's surface is deeply dependent upon the Sun. According to theories of astrophysics, life should not have evolved on Earth at all. The Standard Solar Model

129

predicts that, when Earth was formed, the Sun shone with barely 75% of its present luminosity. Earth's average temperature would have been 15 degrees below zero Celsius, frozen solid. An ice cube planet would reflect most sunlight into Space, causing even colder temperatures. Evolution of life would have been very unlikely.

When we look up from the equations and take a walk before dawn, the variety of life on Earth seems extraordinary. Study of the Earth shows sedimentary rocks produced by liquid water nearly as old as Earth itself. Other geologic markers confirm the presence of rivers and seas 4 billion years ago. Paleontology dates the earliest life forms on Earth at least 3.4 billion and possibly 4 billion years ago. Clearly liquid water and life both existed when the Standard Solar Model says Earth was colder than the movie *Frozen*.

4 billion years ago, while life was gaining a foothold on Earth, heat from meteorite bombardment caused rocks on Mars to melt and crystallize. 15 million years ago another impact caused one small rock to be blown free of Mars and wander through the solar system for eons. 13,000 years ago the little rock impacted Earth in the Allen Hills of Antarctica. It remained in the ice until found by NASA geologist Robbie Score in 1984. For

years meteorite ALH84001 sat on a shelf until geologists realized that it came from Mars.

This author had the honour of working with and being good friends with the late Dr. David McKay of NASA, who passed away in 2013. In 1996 a team led by Dr. McKay found within ALH84001 indications of fossilized life billions of years old. This was the most striking evidence yet found of extraterrestrial life. In 2005 geologist Vicky Hamilton of the University of Hawaii concluded that the rock originated in Valles Marineris, the immense Martian canyon over 5 miles deep. This meteorite indicates that early Mars also had conditions suitable for liquid water and life.

If life existed on Mars billions of years ago, or today, it would be hidden underground. Earth is fortunate to be protected by a magnetic field. Mars' surface, lacking a protective shield, is bathed in radiation from Space. As James Clerk Maxwell showed, light is one form of electromagnetic radiation. Space is so filled with radiation that it poses a major health hazard for our astronauts. Despite the apparent blackness of Space, we live in a Universe of light.

Since the McKay team's discovery about ALH84001, data from spacecraft like Mars Surveyor have found that Mars had liquid water several billion years ago, when the

Standard Solar Model says that Earth and Mars were frozen solid. Photos of Mars show signs of streams, river valleys, and possibly oceans. The fact that life exists to read this book conflicts with the model. This conflict is called the Faint Young Sun Paradox.

At one time scientists thought that huge amounts of carbon dioxide in Earth's early atmosphere raised the temperature. This "greenhouse effect" would have to be precise as a thermostat to keep our climate comfortable. This custom heating was only an inference, for there was no direct evidence for that much CO_2 in Earth's atmosphere. Geologists Robert Rye, Phillip Kuo, and Heinrich Holland have studied the amount of carbon in early Earth's soil. They concluded that the maximum amount of CO_2 that could have been in Earth's atmosphere was far below the amount needed to offset the "Faint Young Sun." Other gases, such as methane, present similar problems. Since both Earth and Mars were warm enough for running water, CO_2 is unlikely to have custom-heated two planets for life.

It should be disturbing when astrophysics predicts that life can't have evolved. Fortunately, the speed of light can help save the Standard Solar Model. The Sun also turns its fuel into energy according to the famous Einstein equation with mc^2. If the speed of light were

132

faster when Earth was formed, both the Sun's luminosity and Earth's temperature would have been higher than originally thought.

Presently Earth is estimated to be 4.6 billion years and the Universe 13.7 billion years old, about 1.5 times its age at time of Earth's formation. Instead of just 75%, the Sun shone with 99% of its present luminosity. The "solar constant" may indeed be nearly constant, allowing life to have evolved on Earth for billions of years.

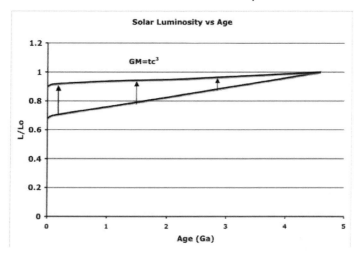

On the graph, the Standard Solar Model of luminosity is an upward curve starting near 75% and rising to reach today's level. When c change is factored in, the luminosity curve becomes nearly level, centred within our comfort zone. Instead of 75%, solar luminosity was almost exactly what we enjoy today. (For cosmologists, Earth's history covers redshifts of 0.3 to 0.)

133

This graph precisely corroborates supernova evidence, distinguishing Theory from "accelerating universe" ideas and models where c is higher only near the Big Bang. If c had not changed by exactly the amounts predicted, we would not be here to talk about it. Sun and supernovae provide two corroborating sets of data from entirely different sources, confirming that c has changed in the amount predicted.

This should be of great comfort in the future, for the Standard Solar Model also predicts that Earth's temperature will continue to rise until the oceans boil. Old models claim that we occupy a fortunate time when Earth is neither freezing nor boiling, just as Earth is the centre of the Universe. Thanks to the speed of light, Earth's temperature will be comfortable for a very long time. Our presence on Earth would not be possible if light had not been slowing down.

Galileo tried timing light with lanterns on distant hilltops, but lacked a good clock. There is an alternative, bouncing light off a hilltop so distant that changes in its speed can be measured with current clocks. Today we have laser lanterns like those in rangefinders. A distant hill has been available since July 20, 1969.

134

Apollo 11 Lunar Laser Ranging Experiment

The Moon

One of the great achievements in human history climaxed when Neil Armstrong and Buzz Aldrin set foot on the Moon. Dr. David McKay was the only scientist in Houston Mission Control when Armstrong walked on the surface. Apollo's immense bounty of science told us, among other things, that the Moon is likely a daughter of Earth torn free in a collision with another body. While

135

studying the Moon as a scientist, I published calculation of a lunar anomaly and the speed of light.

The Lunar Laser Ranging Experiment left behind by Apollo astronauts consists of quartz corner reflectors similar to those on a bicycle. By bouncing laser beams from Earth observatories, astronomers can measure the Moon's distance with great accuracy. Data from the LLRE has told us that the Moon still has a liquid core, verified that Newton's G is indeed constant, and provided one more test of Einstein's General Relativity. LLRE may also have stumbled onto a discovery about light.

The Moon has long been known to be drifting farther from Earth due to tidal forces. As the Moon raises ocean tides on Earth, the tides race ahead following Earth's 24-hour rotation. A tidal bulge tugs the Moon slightly ahead in its orbit, causing the Moon to accelerate. In this way angular momentum is transferred from Earth's rotation to the Moon, causing our length of day to increase and the Moon to spiral outward.

LLRE has measured the Moon's orbital semimajor axis at *384,402 km*. From repeated measurements over 40 years, LLRE reports that distance increasing by *3.82 ± .07 cm/yr*. (The ± .07 cm/yr is a standard deviation.) This rate is known to be "anomalously high". As calculated by Dr. Bruce Bills of NASA with Richard Ray, if the Moon were receding at this speed today, it would have occupied the same place as Earth barely 1.5 billion years ago. Studies of Apollo lunar samples, which I participated in, show conclusively that the Moon has existed separate from Earth over 4.5 billion years.

LLRE is a very precise experiment, checked exhaustively for errors, but it assumes that the speed of light is constant. Fortunately we have ways to measure the Moon's orbit independent of light. One method uses ancient sediments, called tidal rhythmites. Millions of years ago, tides washed up on an ancient shore. Each day the tides left a new layer of sediment, which eventually became fossilized. By studying these tidal rhythmites, geologists can determine the length of the lunar month and the Moon's distance millions of years ago.

The Mansfield sediment of Indiana was formed 310 million years ago, when Indiana was beachfront property. Today the sediment survives in fossilized form, and has been extensively studied by geologists. As reported by Bills and Ray, Mansfield's tidal layers indicate a lunar distance of *375,300 ± 1,900 km*. If we subtract this from today's *384,402 km* and divide by 310 million years, we have the Moon receding at only *2.9 ± 0.6 cm/yr*. Other fossilized sediments also indicate the Moon receding more slowly than LLRE reports.

137

Another method of determining the Moon's recession rate comes courtesy of ancient astronomers. We have eclipse reports going back thousands of years. The track of a total eclipse on Earth's surface is narrow and highly dependent on Earth's rotation rate. If an astronomer in Babylon reported a total eclipse on a certain day in 136 B.C., it provides a precise measurement of how Earth's length of day has changed. In turn this tells how much angular momentum has been transferred to the Moon, and how the lunar orbit has grown.

Dr. F. Richard Stephenson and Leslie Morrison in the United Kingdom have spent many years cataloguing hundreds of historical eclipse observations. Eclipse records covering 2700 years show Earth's length of day changing by only *1.70 ± .05 msec/yr*. This indicates a lunar recession rate of *2.82 ± .08 cm/yr*, in agreement with sedimentary data. LLRE differs by more than 12 standard deviations.

Another way to determine the Moon's recession rate is through computer simulation. Tidal interaction between Earth and Moon is subject to ocean depth, the location of ocean basins, and the movement of continents over time. Simple models of orbital evolution often fail to take these into account. Dr. Eugene Poliakow of the Central Astronomical Observatory in Russia produced a very detailed computer model of Earth-Moon interaction. The simulation has been successfully used to predict today's ocean tides. According to Poliakow's model, lunar recession rate has fluctuated over time. The simulation gives today's rate at *2.91 cm/yr*, agreeing with sedimentary and eclipse data.

If the latter three experiments are given equal weight, their average value would be *2.88 cm/yr*. LLRE's laser light disagrees by *0.94 ± .07 cm/yr*. Mercury's orbit precesses by 5600 arc seconds per century, but a discrepancy of only 43 arc seconds per century was considered proof of general relativity. When LLRE reports the Moon receding 1/3 faster than other experiments, over 12 standard deviations, it is a huge anomaly.

If the speed of light were slowing, time for light to return would increase each year, making the Moon appear to recede faster according to LLRE. $GM=tc^3$ predicts the size of the lunar anomaly to be *0.935 cm/yr*. If the Moon's distance were fixed, LLRE would still report it receding at *0.935 cm/yr* due to a slowing speed of light. This prediction fits the anomaly to less than 1/10 of a standard deviation, a scientific bulls-eye! When prediction fits the data so closely, the chances of the Theory being wrong are vanishingly snall. Like Jupiter's moons and Mercury's orbit, exploration of Earth's Moon gives clues about the Universe and light. As c change would make the Universe appear to accelerate, it would also make the Moon appear to recede faster.

Together these observations show why this is an exciting time to be in science. Signs of a change in the speed of light come from the microwave background, distant galaxies, supernovae, the Sun and Moon. Mysteries that seemed unrelated may be connected. In some cases there are alternate explanations available. It is possible to believe that the early Universe inflated faster than light, if one ignores the contradicting data. Supernova redshifts have caused a variety of speculations, most involving a repulsive "dark" energy. Finally, one can imagine that an unknown force has finely tuned Earth's temperature for our comfort. These explanations share the quality of being invisible and unobserved in the lab. They also share the supposition that we are the centre of the Universe, between extreme heat and cold or between repulsive forces driving expansion. We can believe a Universe of many epicycles, or in $GM = tc^3$. Starting from a first principle is like knowing all the answers in the back of a book.

Sunrise and the International Space Station

ACES: A Better Clock

At JSC we had responsibility to support the International Space Station. ISS orbits a spherical Earth, following laws of gravitation. Newton predicted that an object travelling fast enough it would circle Earth as a satellite. The Station's solar panels convert sunlight into electricity, an application of the photoelectric effect. By exploring over the horizon, we may yet find more clues about the shape of the Universe.

In addition to a more distant hilltop, we can also complete Galileo's experiment with a better clock. The Atomic Clock Ensemble in Space is scheduled be installed aboard ISS in 2016. ACES was designed to be the most accurate clock system ever built. In microgravity an

141

atomic clock is more precise than on Earth, because Earth's gravitational pull does not interfere with the vibration of atoms. An international project led by the European Space Agency, ACES will be orbited aboard a Japanese HTV cargo vehicle and placed by the Station's robotic arm outside the European Columbus module.

One of ACES' chief science goals is to "search for anisotropies in the speed of light." Initial mission length is 18 months, with possible extensions to 3 years or more. At this writing, the International Space Station is funded until at least the year 2020. ACES could potentially measure a change in c smaller than 1 in 10^{10}, or 3.00 cm/sec. Since light would slow this much in 4 years, hopefully atomic clocks will enjoy a long stay aboard ISS. Results may arrive in time for the 100th anniversary of Eddington's eclipse expedition, which helped prove General Relativity. Will the Space Station's atomic clock verify a "c change" in physics? Time will tell!

Evidence that c has changed in time comes from all around, from galaxies, supernovae, the Sun, Moon and possibly the Space Station. Eventually this tide of evidence may be too big to ignore. Many mysteries of the Universe are explained if we open our eyes to light.

Next: Tour of the Universe

142

9

<u>Tour of the Universe</u>

Italy seen from ISS, photos courtesy NASA

If any pleasure approaches that of scientific discovery, it can be found in an aircraft. Flying high above the Earth puts many problems in perspective. Cities at night are ablaze with a million lights. From experience we know that the pools of light are just a hint of the mass below. Between those lights are roads, buildings, and people going about their business. The majority of mass lies hidden in the darkness.

When they have nothing better to do, scientists speculate about multiple universes. There is another Universe, occupying the same Space/Time but hidden from our eyes. Theory predicts, and observations confirm, that the mass we can see is just 4.507034% of the

143

total. To be aware of the other 95.49% is to be a seeing woman among the blind. Here we will take a quick flight around the Universe, searching for what lies in the darkness. We cannot know all the locations of invisible mass, but here are some good places to look.

The White Mountain

Even a journey into Space should begin at home. The Big Island of Hawaii rose from the sea barely one million years ago. The Hawaiian Islands are the product of an ancient volcanic "Hot Spot." This plume of hot lava originates deep within the Earth, near the boundary between core and mantle. The islands sit atop the Pacific Plate, which has been slowly inching across the top of the plume for over 70 million years. The Northwest islands, including Kauai and Midway, are the oldest. The Big Island is the youngest and most volcanically active. The volcanoes Mauna Loa and Kilauea are an exciting place to study the formation of Earth.

This is the Big Island as photographed from a Space Shuttle. To the left is Maui and at top is the Northeast coast including Hilo. The Pacific Trade Winds blow from East to West. Clouds pile up along the coast like waves against a ship's prow, then part to leave an island wake. Hilo is the wettest city in the US, with over 100 inches per year. It is a gardenlike setting surrounded by streams and waterfalls. One reason that Mauna Kea is a good observing site is because the air arrives "clean" without turbulence from another land mass.

Our Big Island's highest point is the dormant volcano Mauna Kea, "White Mountain" in Hawaiian. It's summit, 4,205 metres above sea level, pokes above the clouds and much of the atmosphere's water vapour. From ocean floor to summit Mauna Kea is 10,200 metres high, taller than Mount Everest. Today the summit is studded with astronomer's telescopes, for Mauna Kea is one of the best places on Earth to look at the sky.

The mountaintops of the Big Island have long been considered sacred, with Mauna Kea the most sacred of all. Traditionally only high-ranking chiefs and priests were allowed on the summit. Today the summit is still a place few have visited. The remoteness and altitude make it a difficult place to reach. I waited a number of years to visit the summit because children under 16 are not

145

allowed. Astronomers may spend the night on the summit, but only in a group of at least two. The rarefied air is not for the faint-hearted or anyone with respiratory issues. Most denizens stay at the lodge at the 2700 metre level.

Mauna Kea is home to 12 major telescopes. Recently it has been approved as site of the Thirty Meter Telescope, the largest ever built. The most famous are the twin 10-meter telescopes of the Keck Observatory. The next biggest apertures belong to the 8.3 metre Subaru telescope, the world's largest single-mirror telescope; and the 8.1 metre Gemini North telescope, which can see in both visible and infrared light. The Canada-France Hawaii Telescope (CFHT) also looks into the infrared sky. Part of the summit is named "Millimeter Valley" for observatories that peer into the ultraviolet and radio frequencies. These millimeter waves allow astronomers to peer through some of the dust that fills our galaxy.

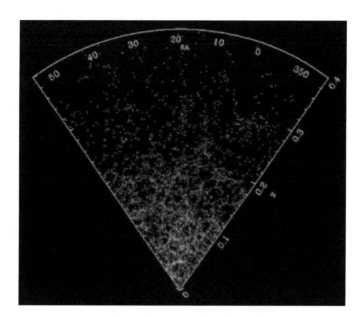

Astronomer Margaret Geller and my friend the late John Huchra have been mapping galaxies since 1985. Today the Sloan Digital Sky Survey has mapped a slice of the visible Universe. Galaxies and clusters form enormous sheets stretching across millions of light-years. These great walls are boundaries of vast bubbles containing most of the Universe's volume.

A fish in the Barrier Reef avoids dark holes in the coral, because they could hide something that could eat her! One hopes that humans are more intelligent than fish. It would be foolish to think that humans know everything in the Universe, or to assume that the "voids" are empty. Something may indeed be in those dark places, something hungrier and more massive than humans have imagined.

147

Theory predicts that quantum fluctuations near the Big Bang could have created singularities of almost unlimited size. The largest would be true Black Holes devouring everything within reach. In the early Universe they would have cleared whole regions of matter. Models predict that about 68% of mass in the Universe would end up in such regions. Since the Cosmic Microwave Background dates from 380,000 years after the Big Bang, this majority of mass would not show up in the CMB. The fragile spaces in between, balanced in a tug-of-war between masses, would form sheets where galaxies could form. Magnified by expansion of the Universe, the picture would look exactly like what Sloan has seen.

Despite their immense size, ultra-massive Black Holes would be very difficult to detect. Their intense gravity would devour any light or radiation. They would, however, produce magnetic fields that could be detected. Astronomers have found such powerful magnetic fields, the source of which has been a mystery. Black Holes can also be located through gravity. As this is being written, one such object is already drawing us inexorably toward it.

The Milky Way is within a Local Group of about 30 galaxies. This group is part of a Local Supercluster containing dozens of clusters spread over 100 million light-years. At the centre of this is the Virgo Group, which contains thousands of galaxies. Everything in this vast community is being inexorably drawn at 600 km/second toward a mysterious mass in the direction of cluster Abell 3627. This immense mass, about 10^{16} times the mass of our Sun, has never been seen. Its discoverers have dubbed it The Great Attractor. This object is not alone; there is evidence of another Attractor hidden beyond it. We only know of the Great Attractor's existence because it influences our galaxy directly; the Universe could contain many others.

During 2014 astronomers at University of Hawaii, using a new technique for determining distances, reported that our Virgo Cluster is just a small part of an enormous supercluster This immense structure has been named Laniakea, Hawaiian for "immense heaven". Laniakea contains 100,000 times the mass of our Milky Way galaxy and spans 500 million light years. The galaxies within Laniakea, including ours, are slowly being drawn toward the unseen Great Attractor.

There are other ways to detect what cannot be seen. This photo is from the Hubble Telescope Deep Field. General Relativity predicts that gravity bends light rays. Abell 2218 is a cluster of galaxies, each containing billions of stars surrounding a massive Black Hole. Around the edges we see galaxies shifted into red. These galaxies are not located around the cluster but behind it, 5-10 times more distant. Their light has been bent by gravity, traveling around the cluster to reach us! The images of galaxies have been stretched into arcs, as if seen through an immense lens. This phenomenon is called gravitational lensing. This bending of light is a spectacular confirmation of General Relativity.

With the aid of computers, we can use gravitational lensing to see where the mass really lies. The cluster's density peaks at the galaxies, but is even higher toward the centre. Most of the cluster's mass is dark. Something in the centre holds the cluster together and bends the light from more distant objects. Whatever lies there must be far more massive than the galaxies, and concentrated in a small region. We could be seeing the lair of an ultra-massive Black Hole. This object did not form a galaxy because it swallows all light and matter in its vicinity.

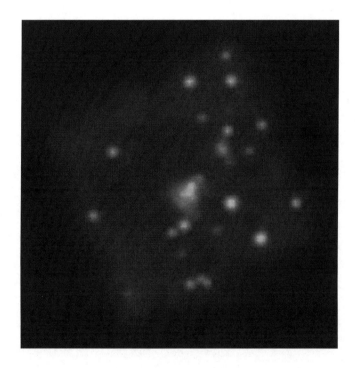

The European Space Agency's XMM-Newton spacecraft took this X-Ray photo of Cluster RXJ0847.2+3449. We see this cluster of galaxies as it was

151

about 7 billion years ago, when the Universe was half its present age. A team led by Alain Blanchard of the Midi-Pyrenees Observatory in France has found many such clusters. They help confirm whether "dark energy" really exists.

If a repulsive energy filled the Universe, 5 billion years ago it would have begun to dominate. Formation of galaxy clusters would have slowed as repulsion overcame gravity. We would see more clusters in the past than exist today. Astronomers have found significantly fewer clusters in the past, telling us that clusters are still forming. Rather than tearing itself apart with "dark energy," the Universe continues to form more complex structures.

Results from XMM-Newton, which are verified by the Chandra X-Ray Observatory, show that gravitating "dark" mass is four times as abundant as previously thought. Instead of just 24%, it comprises the 95.49% of the Universe that is not baryons. It is by far the majority of mass in the Universe, where our matter is just a footnote. There is no need for a repulsive "dark energy" to fill the rest. We are the lights hinting at something else in the darkness.

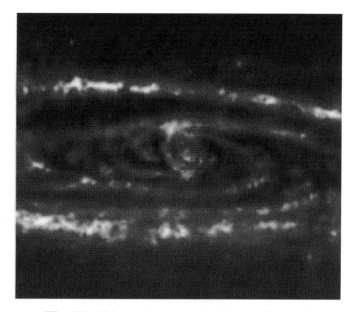

The Hubble expansion does not affect all galaxies equally. Andromeda, the nearest galaxy, is moving toward us. The Spitzer Space Telescope made this infrared image. 200 million years ago galaxy M32 collided with Andromeda, leaving huge ripples in Andromeda's disk. In another 5-10 billion years Andromeda and our Milky Way will collide. The two galaxies do not contain enough mass to cause this attraction. There is invisible mass between them, estimated to be ten times the mass of a galaxy.

Andromeda's core is a Black Hole with about a million times our Sun's mass. Not big enough to swallow everything, it formed the seed around which the galaxy formed. Caltech scientist Fritz Zwicky first suspected that "dark matter" surrounds and fills the galaxies. By measuring the rotation rates of stars at various distances from galactic centres, astronomer Vera Rubin concluded that galaxies contain far more than meets the eye. A halo of invisible mass surrounds each galaxy.

Cosmologists believe that the Big Bang created billions of tiny Black Holes. Attracted by gravity to larger objects, they would form haloes around the galaxies. Exactly like Zwicky's "dark matter," they would exert gravitational influence but emit no light. Models predict this mass at 23.87% of the total, a figure confirmed by WMAP. The term "dark matter" may itself be misleading. These objects would have formed directly from quantum fluctuations and have never been matter.

Globular cluster Messier 80 contains hundreds of thousands of stars and orbits around our Milky Way. Astronomer Harold Shapley used observations of these objects to locate the Milky Way's centre. A globular cluster contains some of the oldest stars in our galaxy. Astronomers did not know how the globular clusters formed so early, or what holds them together. One way to form them would be from a medium-sized Black Hole orbiting the Milky Way. Since this Black Hole formed primordially, it would have been a magnet for early star formation.

Star formation is related to another question. If our galaxy is orbited by billions of Black Holes, have they not collided with the galaxy's disk? Black Holes striking gas clouds in our galaxy would surely produce effects we could see. They may have collided with the galaxy billions of times, and we could be the result.

One of the most spectacular pictures from the Hubble Space Telescope is of the Eagle Nebula. Here we see the process of star formation. The nebula is a gas cloud containing raw ingredients of stars. The immense pillars are cocoons containing infant stars. Somehow the

156

highly diffuse interstellar gas is collapsing into a point so hot and dense that nuclear fusion is ignited. During this process, something prevents that heat from dissipating the cloud and ending the birth.

One way to produce this formation would be if a tiny Black Hole, or many, collided with the cloud. Small singularities would naturally gather materiel around them, but would not be big enough to suck up the whole cloud. They would leave behind pillars of gas like pellets fired through cotton candy. At their centres, gravity and Hawking radiation would produce the conditions for stars to ignite. The heat produced would balance gravity's inward pull so that the stars burned steadily. The Black Holes may still be in the stars' cores, quietly contributing to their power output.

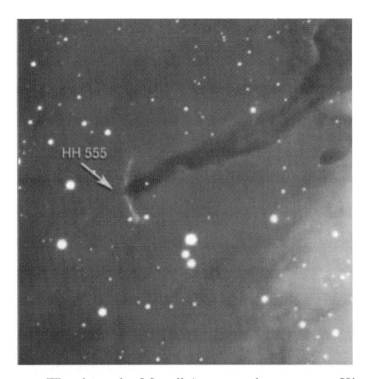

Thanks to the Mayall 4-meter telescope atop Kitt
Peak, this is HH 555, a Herbig-Haro object in the Pelican
Nebula. At the pillar's tip gas is drawn into a disk, which
will form a baby star. At the top and bottom are two
bright jets, guided by magnetic fields from the core. Twin
jets and a magnetic field are telltale signatures of a hidden
Black Hole.

Theories of the Sun have advanced over time. As
late as the 1920's most astronomers would lecture that our
Sun was made of iron, and glowed in the sky like a hot
poker. Only a young astronomer named Cecilia Payne
suggested that the Sun's spectral lines could be interpreted
as hydrogen. Possibly because Payne was a woman, her

158

idea was roundly dismissed. Though she had completed studies at Cambridge, the University did not award degrees to women. The equations of nuclear fusion were still being worked out, and most scientists doubted that Black Holes existed. Eventually the young woman was vindicated. As our knowledge of physics advances, so must theories of the Sun.

The reader may guess where this line of thought is leading. This is a Sunset seen from the East coast of the Big Island. A Black Hole could conceivably exist in the second last place we would think to find one, inside the Sun! No physical law prevents this; most laws of astrophysics would remain unchanged. A Black Hole would feel right at home in the radiation and pressure of a stellar interior. If it had memory, it would be reminded of the conditions, which accompanied its birth near the Big Bang. Presence of a tiny singularity, which is predicted by cosmology, would have ignited the nuclear reactions within a star. The primary power source of stars would then be fusion, occurring in the immense heat of stellar cores. The Black Hole's rotation would cause the star's inner layers to rotate faster, contributing to a magnetic field. Indeed astronomers now know that our Sun's core rotates faster than the outer layers due to some mysterious influence. A Black Hole could be literally in front of our face each morning.

Niels Bohr is best known for describing the heart of an atom, but he also turned his attention toward stars. Bohr's atomic model is based upon a miniature Solar System. In the early part of the 20th century many old ideas of science were questioned. Bohr would say that "Your theory is crazy, but it's not crazy enough to be true." In work that was never published, but related to his friend George Gamow, Bohr hypothesised that a star's heart is made of an unknown material radiating at a constant temperature. The region surrounding this core would be composed of hot plasma. If temperature of the core were greater than that of the plasma boundary, an equilibrium would be reached. The star would maintain a

160

constant luminosity.

Bohr's hypothesis required a material radiating at a constant temperature, something unknown in the 1930's. Though Black Holes are predicted by Relativity, even Einstein himself didn't believe they exist. In 1974 Stephen Hawking reached the amazing conclusion that they radiate at a temperature depending upon their mass. For many Black Holes this temperature would be nearly constant. Connecting Bohr with Hawking, a Black Hole at a star's heart would be just the thing to maintain a constant luminosity.

Only occasionally would Black Holes within stars reveal themselves. In their twilight years, the largest stars would consume all their fuel until fusion abruptly stopped. The equilibrium between outward pressure and gravity would end, causing a catastrophic collapse. A star's mass suddenly falling into a Black Hole would produce an immense explosion, a supernova. An exploding star is more than a dot in an astronomer's photo. These titanic explosions would end all life not just in their own solar system but around nearby stars too. A supernova nearby in our galaxy would render Earth uninhabitable, a fate we have so far been lucky to avoid.

A supernova leaves in its aftermath a neutron star, composed of matter so dense that a spoonful would weigh

161

a ton. Some neutron stars are pulsars, emitting spinning beacons of radiation like a lighthouse. Cause of these jets has been another mystery. The axes of the jets do not coincide with the neutron stars' axis of rotation, indicating that something else inside the neutron star creates the jets. The pulsar's jets are exactly like the twin jets from a Black Hole. Other neutron stars are magnetars, objects with magnetic fields so powerful that they defy Maxwell's equations to describe them. Once again, jets of radiation and a magnetic field are telltale signatures of a hidden Black Hole. It would rotate independently inside a neutron star, producing both a spinning beacon and magnetic field.

A journey ends with a return home. We began at our own planet, by seeing how our early pictures of Earth evolved. From there we have visited the solar system, the stars, the galaxies and even the beginning of the Universe. Now we can see our home with new eyes. One of the biggest surprises of cosmology may wait beneath our feet. It was created shortly after the Big Bang and has been waiting since before life began for our discovery.   Earth contains enough mysteries for another book.

Next: A Theory for Everything?

10

<u>A Theory For Everything?</u>

Starting from a child's first explorations, we have wandered from the beginning of the Universe and back to our home planet. Science has discovered that similar rules apply to the whole Universe. In many ways, every part of the Universe resembles every other. Today we may be able to determine the size and shape of the Universe. Our seemingly complicated Universe may be described in a simple equation.

Our position today is like that of humans thousands of years ago, standing on the shore wondering about Earth's shape. Some said that it was flat, others looked at evidence and concluded Earth was spherical. Eventually the curved theory won out. Today we are barely able to find evidence for curvature on the Universe. With more data, our children will understand its size, shape, and the simple principles it is based upon.

A spherical Universe, resembling Earth or a mother's womb, appeals to our sense of aesthetics. The simplicity of spherical shapes appealed to Pythagoras, Pascal, Einstein and even Edgar Allen Poe. If someone thinks that the Universe is spherical, they are in very good

163

company. Today their intuition has been borne out by mathematics.

Topology is the study of surfaces. In topology, shapes can be stretched like dough and still be mathematically similar. A sphere is similar to a cube in that one shape can be stretched into the other without tearing any holes. A donut and a teacup are similar in that they each have one hole.

From tiny cells to stars, 3-demensional objects tend to form spheres. Henri Poincare was the father of modern topology. In 1904 he conjectured that the most natural form for a 3-dimensional space is the surface of a sphere in 4 dimensions. Poincare could not prove his conjecture, and it remained unproven for a century.

Our astronauts experience a simple form of Poincare's Conjecture every time they venture outside on EVA. 2-dimensional sheets of fabric and rubber have been made into human-shaped spacesuits. Upon emerging into the larger world of Space, pressure tends to balloon the spacesuit into the shape of a sphere. The Spacesuit designers and the astronauts must fight this tendency to form spheres, making movement during EVA very difficult.

Grigorii Perelman was born in St. Petersburg, then called Leningrad in the former Soviet Union. His father,

164

an electrical engineer, encouraged an interest in science and gave his son many books. His mother, a math teacher, took Grigorii to operas and encouraged him to enjoy music. Perelman showed an early talent for maths, attending a specialized school for mathematics and physics before entering Leningrad University at age 16. After graduation, he took a post at the Steklov Institute of the USSR Academy of Sciences. After the collapse of the Soviet Union, Perelman came to the US and found work at American universities. In the United States he was exposed to new worlds of mathematics, including Poincare's mystery.

Mathematician Richard Hamilton earlier had proposed a solution based upon Ricci flow. This flow insures that the Ricci curvature tensor, important in General Relativity, is always positive. Heat introduced into a tub of cold water will diffuse out until the water has a uniform temperature. Hamilton suggested that curvature could also diffuse out until a surface of equal curvature, a sphere, was achieved. Proving this in 3 dimensions seemed an insurmountable problem. Perelman met Hamilton while in the US, then returned to Russia in 1995. For the next 7 years he would focus almost exclusively on proving the Poincare Conjecture.

By the turn of the millennium, technology was at the point where scientists could work in isolation and still communicate ideas. Scientists from Aristarchus to Galileo wrote their own books to spread ideas. Einstein first published in journals, which had to pass review by suspicious editors. In November 2002 Perelman began posting outlines of a proposed proof on the internet. Though some mathematicians realised he was on to something, parts of the proof were incomplete. By Spring 2006 the proof was complete, and the world realised what had been accomplished.

Perelman's achievement made him eligible for multiple maths prizes. Famously he has not bothered to collect them. Like many great minds, he prefers to work in privacy. Perelman considers himself retired from the math field, and takes his greatest pleasure in the opera. He prefers the inexpensive seats high in the opera house. Away from distractions, he can listen for the harmonies in singer's voices. Like Pythagoras, Perelman enjoys both music and maths.

The Poincare Conjecture has applications to our Universe. It implies that the only possible shape for a 3-dimensional Universe is the surface of a 4-dimensional sphere. Einstein calculated that if density of the Universe were great enough, gravity would curve it into such a sphere. This shape can be maintained without collapsing if there is a certain density. That density is not "critical" but the stable density. Mathematically this density is an "attractor," pointing to just one possible Universe. If the Universe were created from nothing, then mass would be created to push the Universe toward this density.

Since it is possible to describe the Universe with one equation, others have breathlessly speculated about a "theory of everything." Describing even the form if such a theory would be very difficult. Einstein, in his later years of fame, worked fruitlessly to find a "unified field theory" combining gravity and electromagnetism. Many good minds have been wasted trying to unify General Relativity with Quantum Mechanics. It may be a waste of time trying to create an all-encompassing theory of everything. Physicists would oppose such a theory, for it would put many of them out of work.

Small links between General Relativity and Quantum Mechanics already exist. Hawking's prediction of Black Hole radiation uses both. We have seen that the speed of light c, which is a part of Relativity, and the Planck value h of Quantum Mechanics may be linked. Instead of an overarching theory of both, QM and GR may be stitched together from small links.

As promised, there is only one equation in this book. However, an equation may have more than one form. Our units, like meters and seconds, are based on arbitrary values. An alien species would use entirely different units. Max Planck in 1899 suggested a system of units based on combinations of his value h, the speed of light c, and Newton's gravitational constant G. The "Planck units" of mass, length and time are made up of combinations of these values. These tiny units may have significance for our big Universe.

The only equation in this book, when combined with an expression for the Universe's scale R, becomes even more simple:

$$M = R = t$$

Mass M of the Universe, about 10^{60} Planck masses, equals its radius R (10^{60} Planck radii) equals its age t. We may call this The Equation of the Universe.

The number 1 followed by 60 zeroes, like the Universe, is very large. Arthur Eddington, whose eclipse expedition tested Einstein's General Relativity, thought that large numbers like these were significant. In his later

168

years he used them as the basis for a "fundamental theory" encompassing both the very large and very small. Paul Dirac, who predicted antimatter, hypothesised that the electromagnetic force, gravitation, size of the Universe and size of an atom were all related. Dirac made the prediction that the gravitational constant G changed over time. Later experiments have not supported his prediction. However, Dirac's large-number problem is also solved when $GM=tc^3$.

Nature contains many values and parameters. These include the masses of the proton and electron, their electric charge, and other measured quantities. Numbers without dimension include the fine-structure value α and various ratios found in nature. The standard model of particle physics contains many "free parameters" that can be measured but not yet predicted. This book has shown how one important value, the speed of light c, can be calculated at different times. In the future other values and supposed constants may be calculated from pure mathematics. The Universe large and small contains many mysteries yet to be solved.

The desire to understand the Universe is part of humanity. It is a natural extension from crawling, walking and climbing. To fly is to reach and explore far more places. Every human has at one time or another

169

dreamed about flying. Consciously we have imagined angels and superheroes that resemble us but can fly. Though humans have no wings, we are given minds that can dream and organise. The ability to dream has enabled us to build flying machines and spacecraft. Humans are the only species on Earth that can leave the planet.

The Universe continues to expand and grow more complex, as it has for billions of years. It will do so for countless hundreds of billions of years more. Just as a time near the Big Bang seems like a beginning to us, our time will someday seem like a beginning.

Like children first opening our eyes to light, we are only beginning to understand the Universe.

L. RIOFRIO
Houston, Texas

170

11

<u>Afterword: The Year of Light</u>

BICEP2 on left and South Pole Telescope on right.

The year leading up to February 19, Copernicus' 542nd birthday, has been very exciting! Not only was this book published, but a decades-old paradigm of the early Universe made its last inflated bubble before bursting. Tonight we are in the midst of celebration, for 2015 is The Year of Light.

Antarctica is Earth's last frontier. Aristotle, who was fascinated by balance, thought that a Southern land, Anti-Arktos, must exist to balance the landmasses in the North, which were called Arktos. For centuries it was simply Terra Australis Incognita, a mystery. Captain James Cook in the 18th century tried and failed to find the Southern Continent. Over decades Antarctica was discovered, explored, and finally settled by scientists. Today you may join cruises and scientific expeditions to the bottom of the world.

The Southern continent is also a new frontier for science. Many meteorites have been preserved in the ice. In the Allan Hills Robbie Score picked up ALH84001, which my NASA colleagues found to be from Mars. The little rock, along with other meteorites from Mars, still show tantalising signs of ancient life. Antarctica could be the site of major scientific discoveries.

Like Mauna Kea, the South Pole is an excellent site for astronomy. In the higher elevations, the cold polar air is very free of moisture. Lack of water vapour allows light to reach our telescopes unclouded. The South Pole Telescope is located at an altitude of 2800 meters. Sharing the SPT's home is a smaller telescope with the odd name of BICEP2, built for the sole purpose of finding

172

gravitational waves.

We sail on an ocean filled with water waves. Gravitational waves are thought to be a prediction of Einstein's General Relativity. Despite decades of searching by many experiments, no one has ever found them. In the late 1960's scientist and former Naval Officer Joseph Weber made headlines by claiming to find the waves, but no one could reproduce his results. Weber would find patterns in random data, seeing the result he hoped to find.

For decades the paradigm called "inflation" has dominated study of the early Universe, crowding out other ideas. Inflation claims that the sky is filled with a sea of gravitational waves. The Wilkinson Microwave Anisotropy Probe found no waves, which ruled out some versions of inflation. Stephen Hawking made a bet that the PLANCK spacecraft would see gravitational waves, but PLANCK found none. Many experiments, including the giant LIGO interferometer and the South Pole Telescope, have been in a race to find waves. Einstein also said that insanity is doing the same thing repeatedly while expecting a different result.

BICEP2 was intended to see signs of waves in the microwave background. This is a complex subject, but waves were thought to cause polarization of the radiation, which computers could then ferret out. Like water waves filling the sea, inflation was thought to fill the sky with a pattern called "B-modes". The collaboration that built BICEP2 included researchers from the Harvard-Smithsonian Center for Astrophysics and Stanford University. The neighbours at the South Pole Telescope had found B-mode polarization in 2013, but attributed it to foreground interference. Spending several cold Winters at the South Pole, the BICEP2 collaborators saw the result they hoped to find.

Einstein's birthday is easy as Pi to remember, 3.14. On Friday March 14, 2014 Paul Steinhardt, along with his colleagues Anna Ijjas and Abraham Loeb, published a paper online concluding that inflation was not supported by observations. Steinhardt, the Albert Einstein Professor of Physics at Princeton, created one of the earliest inflation models in 1983. In 2002 Steinhardt, Alan Guth and Andre Linde shared the Paul Dirac Prize for their work. Since then Steinhardt has grown tired of inflation, citing its many problems. The new paper included data from WMAP, the Atacama Cosmology Telescope in Chile, and the PLANCK spacecraft. That Friday Steinhardt's paper had already been upstaged. Rumours had spread online that a major announcement would be made Monday.

On March 17 in a press conference at Harvard University, BICEP2 announced detection of gravitational waves from the beginning of the Universe. The four main BICEP2 collaborators appeared before a picture of an inflated Universe, with Alan Guth and Andre Linde proudly watching in the audience. The "smoking gun" was front-page news in *The New York Times*, *Nature* and around the world. It was called "The greatest discovery of the century".

The next day, in a packed lecture hall at MIT, Guth gave a triumphant talk. The press was predicting the most famous scientific Prize for inflation, the gold medal awarded in Stockholm. Guth himself claimed that the discovery was "definitely" worth the Prize. One newspaper asked, who would get the Prize? Success has many fathers, and hundreds of men had sired inflation theories. Even Stephen Hawking claimed to press that he had organized a conference where inflation was midwifed

Stanford University released a viral video of a BICEP2 collaborator, wearing a backpack and walking with his hands in his pockets, visiting Andrei Linde's home near Stanford to give him the news. Linde's wife Renata Kallosh, who is also a physicist, answered the door. The visitor poured champagne and also told the world of the "smoking gun". The Stanford video was seen by over 2 million viewers.

Scientists usually wait until a paper is published to announce results. The process of peer-review is arduous, especially for new ideas, but it can serve as a check on premature findings. BICEP2 posted their results online without getting them reviewed or published in a journal. The "smoking gun" was not checked.

From the beginning saner heads questioned BICEP2 noise. Analysis of the microwave background is affected by processing of data, which brings bias. The so-called B-modes are very difficult to see. They must be "found" using statistics to separate them from random noise. Dust grains in interstellar space can also cause B-modes, as background light is polarized while passing through dust.

Even if B-modes come from gravitational waves, other phenomena besides inflation could make waves. Colliding stars and Black Holes are thought to cause gravitational waves, though none have been directly detected. Many other experiments, even spacecraft like WMAP and PLANCK, have searched for waves and found none. Despite the hype, a few thought that the gun was not smoking.

On May 12 growing doubts were summarised by physicist Adam Falkowski in his blog *Resonaances*. The biggest problem was, quite simply, dust. Our galaxy, like an old home, is full of dust. Before radio telescopes, humans could not even see the galaxy's full extent because of dust. Though today we know that the galaxy's centre is in the constellation Sagittarius, we can't directly see there. Some of this dust has condensed into stars and solar systems, but much remains. The effects of dust must be subtracted whenever we look at the microwave background.

BICEP2 had tried to avoid the problem by aiming at the Southern Hole, a patch of sky thought to be free of galactic dust. The collaborators were not sure where the dust was located in the sky. The best dust maps had been produced by the PLANCK spacecraft, but that data had not been released. An early map had been shown at a conference talk by a PLANCK scientist, then posted on the conference website. The BICEP2 collaborators used the map without permission, and "scraped" the dust data from it.

David Spergel, also of Princeton University, is a leading expert on the Cosmic Microwave Background. He had been one of the chief scientists of the WMAP spacecraft, which found no signs of gravitational waves. While riding a train to lecture in New York, he read the paper that BICEP2 posted online and quickly had doubts, which his colleague Randolph Flauger summarised in a talk on May 16. On May 28 Spergel, Flauger and Colin Hill published their misgivings online .

When challenged, BICEP2 stuck to their smoking guns. One co-leader of the group insisted, "We stand by our paper", and would not revise or retract it. When asked if they admitted to a mistake, the principal investigator claimed, "We've done no such thing". Another co-leader of BICEP2 insisted that their paper was "Certainly not being retracted."

Inflation's bubble continued to grow. On May 31 Guth and Linde (but not Steinhardt) were named winners of the Kavli Cosmology Prize. They travelled to Oslo to collect a treasure of Norwegian Kroners--many thought that the next stop would be Stockholm. BICEP2's claims were very rewarding for inflation's proponents.

On June 3 *Nature*, which had earlier trumpeted the "smoking gun," published an article about the growing doubts. In the same issue of *Nature* Paul Steinhardt was given space to write a blistering critique of the whole inflated idea. He was shocked to hear inflation's proponents say that if there were no gravitational waves, inflation would not be ruled out. They could simply imagine an alternate inflated universe with undetectable waves. He concluded that inflation is "fundamentally untestable and scientifically meaningless". Steinhardt is

178

currently working on an alternative to inflation.

The BICEP2 paper was finally published in *Physical Review Letters* on June 29. The collaborators had revised their conclusions, saying that "B-modes" could be caused by dust. Unheeded, the next day *National Geographic* published an idolizing profile of Alan Guth. The article noted that in front of Guth's messy office at MIT is an empty wood and glass display case, already prepared for the medal from Stockholm. On August 1 Guth and Linde, as winners of the Kavli Prize, visited the White House to meet the President. The September 16 cover of *Scientific American* hailed a "Beacon from the Big Bang."

The cover stories may have been premature, appearing just days before a paper from the PLANCK group. Data from the spacecraft had found no evidence of gravitational waves. Furthermore, the paper concluded that the signal from BICEP2 was probably cosmic dust. The PLANCK paper was devastating. Adam Falkowski wrote that Guth and Linde could not even dream of winning the big Prize. Inflation's bubble had grown, shown its face to the world, and burst.

In the aftermath PLANCK scientists agree to share their data with the BICEP2 collaborators, checking the results. In December *Nature* named David Spergel one of 10 scientists who made a difference in 2014. None of the collaborators of BICEP2 or inflation were mentioned. As 2014 ended, most scientists doubted that BICEP2 had seen anything.

On January 30, 2015 the new PLANCK-BICEP2 marriage released a joint paper. It concluded that the "gravitational waves" were almost certainly dust. Having been misled the year before, the press concluded "Gravitational waves discovery now officially dead". Other headlines were even more critical, saying, "Evidence for cosmic inflation theory bites the dust". As before, there was no evidence that "inflation" ever happened. The paradigm that had dominated cosmology for 35 years was had breathed its last.

Inflation was conjured to explain two problems of the Big Bang. One question is why large parts of the sky appear to be at the same temperature. The other puzzle is why the density is close to a "critical" value, neither too dense nor underdense. Inflation solves neither of these problems, as it must be precisely fine-tuned to produce the Universe we observe today.

Speculations about inflation have been a useful step, but have never been a real theory. Even those who religiously worship inflation must ask: What would cause the Universe expand faster than today's speed of light? Perhaps light was faster. As I write this we celebrate light, for this is a special year.

On December 20 a resolution of the UN General Assembly officially declared 2015 The Year of Light. 1000 years ago in Baghdad, Ibn Al-Haytham was writing the first treatise about light. 200 years ago Auguste Fresnel studied the refraction of light. 150 years ago James Clerk Maxwell wrote equations showing that light was a form of electromagnetic radiation. 100 years ago Einstein published his General Relativity. 50 years ago the Cosmic Microwave Background, was discovered. The Year of Light is being coordinated through UNESCO. Around the world many celebrations, lectures, fireworks and light shows commemorate this year.

The cover of this book shows the Sun rising over the Pacific, seen from the International Space Station. In 2016 the Atomic Clock Ensemble in Space is scheduled be launched to ISS. Galileo tried to measure light's speed with lanterns on distant hilltops, and dreamed of an accurate clock. In ACES we may soon have the most accurate timepiece ever built. One of the clock ensemble's principal goals is to search for anisotropies, or changes in c. We may soon see another discovery about light, in a year of ACES.

Many believe that there is trouble with physics. There

have been precious few predictions in the past 35 years that led to discoveries. Physics has been mired by speculations of inflation and other ideas that are "not even wrong." Light solves the problems that inflation and "dark" energy were imagined to solve. It would open the door to many other discoveries.

The Universe began in a burst of light. Astronomers celebrate First Light when a new telescope is opened to the Universe. A baby experiences First Light when her eyes first open. Light brings discovery while "dark" energies spring from fear and ignorance. Humans may open their eyes to light.

Bibliography

Chapter 2
Ptolemy, *Geographia*

Copernicus, Nicholas, *De Revolutionibus Orbium Coelestium*, 1543

Galilei, G., *The Starry Messenger*, 1610

Galilei, G., *Dialogue of the Two World Systems*, 1632

Chapter 3
Newton, Isaac, *Philosophia Naturalis Principia Mathematica*, 1686

Newton, Isaac, *Opticks*, 1704

Thomson, W. and Tait, P.G. *Natural History*, Vol. 1, No. 403, 1874

Chapter 4
Einstein, Albert, "On a Heuristic Viewpoint Concerning the Production and Transformation of Light," *Annalen der Physik* Vol. 17, No. 6, 1905

Einstein, Albert, "On the Electrodynamics of Moving Bodies," *Annalen der Physik*, Vol. 17, No. 10, 1905

Zeeman, E.C., "Causality Implies the Lorentz Group," *Journal of Mathematical Physics*, Vol. 5, No. 490, 1964

Chapter 5
Einstein, Albert, *Relativity: The Special and General Theory*, Princeton University Press, 1961

Poe, Edgar Allen, *Eureka*, 1848

De Broglie, Louis, "Researches on the quantum theory," *Annalen der Physik*, Vol. 10, No. 3, 1925

Chapter 7
Riofrio, L., "GM=tc^3 Space/Time Explanation of Supernova Data," *Beyond Einstein*, 2004, http://www-conf.slac.stanford.edu/einstein/talks/aspauthor2004_3.pdf

Chapter 8
Bills, B.G. and Ray, R.D., "Lunar Orbital Evolution: A Synthesis of Recent Results," *Geophysical Research Letters*, Vol. 26, 1999

Stephenson, F.R. and Morrison, L.V., "Long-term fluctuations in the Earth's rotation," *Philosophical Transactions of Royal Society*, 1997

Poliakow, Eugene, "Numerical Modelling of the paleotidal evolution of the Earth-Moon system," *Proceedings of Intl Astronomical Union 197*, 2004

Riofrio, L., "Calculation of lunar orbit anomaly," *Planetary Science*, Vol. 1, No. 1, 2012, http://www.planetary-science.com/content/pdf/2191-2521-1-1.pdf

Chapter 9
Geller, Margaret, and Huchra, John, "Mapping the Universe," *Science* Vol. 246, No. 4932, 1989

Blanchard, Alain et al., "An alternative to the cosmological concordance model," *Astronomy and Astrophysics*, Vol. 412, No. 1, 2003

Chapter 10
Perelman, G., "Finite extinction time for the solutions to the Ricci flow on certain three-manifolds," 2003 http://arxiv.org/abs/math.DG/0307245

Dirac, Paul, "The Cosmological Constants," *Nature* Vol. 139, No. 3512, 1937

Eddington, Arthur, *Fundamental Theory*, Cambridge University Press, 1946

Memories
Day 1

Memories

Day__

Memories
Day__

_____
_____
_____
_____
_____
_____
_____
_____
_____
_____
_____
_____
_____
_____
_____
_____
_____

Memories

Day__

Memories
Day__

Memories
Day__

192

Memories

Day__

_____
_____
_____
_____
_____
_____
_____
_____
_____
_____
_____
_____
_____
_____
_____
_____
_____
_____

Memories

Day__

Memories

Day__

Memories

Day__

Memories
Day__

Memories
Day__

Memories
Day__

Memories

Day__

Memories

Day__

Memories

Day__

Photo

Day__

Made in the USA
Lexington, KY
14 February 2018